T0214748

Problem Books in Mathematics

Series Editor:

Peter Winkler
Department of Mathematics
Dartmouth College
Hanover, NH 03755
USA

More information about this series at http://www.springer.com/series/714

Andy Liu • George Sicherman
Takayuki Yoshigahara

The Puzzles of Nobuyuki Yoshigahara

Springer

Andy Liu
Edmonton, AB, Canada

George Sicherman
Wayside, NJ, USA

Takayuki Yoshigahara
Tokyo, Japan

All puzzles described in this book were originally conceived by Nobuyuki Yoshigahara (deceased).

ISSN 0941-3502   ISSN 2197-8506   (electronic)
Problem Books in Mathematics
ISBN 978-3-030-62895-6   ISBN 978-3-030-62896-3   (eBook)
https://doi.org/10.1007/978-3-030-62896-3

Mathematics Subject Classification (2020): 00A07, 97D50, 97U40

This Springer imprint is published by the registered company Springer Nature Switzerland AG
The registered company address is: Gewerbestrasse 11, 6330 Cham, Switzerland

Dedicated to the memory of

John Conway,

Martin Gardner,

Solomon Golomb,

Richard Guy,

Klaus Peters,

Raymond Smullyan

and

Nob Yoshigahara

Preface

My fondest memory was sitting at my father's knees as he designed mechanical puzzles in the small craft shop at home. Apart from a creative mind, he also had dexterous hands. Many of his puzzles were marketed, but I had the special privilege of playing with a huge number of puzzles he was developing but did not eventually release.

My father was thinking of puzzles every day. He always carried a notepad to jot down inspirations of the moment. Although he had a puzzle column in several magazines, he was always able to meet their deadlines.

My father was intoxicated with both puzzles and alcohol. He could spend an infinite amount of time on puzzles, but a small amount of alcohol was enough to make him happy. He was a frail person, and had battled and survived cancer of the stomach at age fifty-six.

Over the years, my father published more than eighty puzzle books in Japanese. "Puzzle 101: A Puzzlemaster's Challenge", published by *A K Peters*, was his only book in English. He was preparing a second volume when he passed away.

I was very pleased when Andy Liu contacted me about publishing more of my father's puzzles in English. It would fulfill one of his dying wishes. Accordingly, I supplied Andy with a selection of material from my father's personal file. I hope that a book will come of it that would rekindle in the west interest in my father's puzzles.

To conclude this brief narrative, I now quote one of my father's favorite puzzles. The numbers in the diagram below are put together according to a certain rule. What number should go into the blank circle?

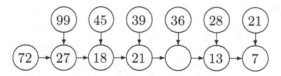

"I have solved your puzzle, Nob," many people told him. "However, I must tell you that there is a typo in it. The number in the last circle should be 8, not 7."

My father just shook his head and smiled.

Takayuki Yoshigahara,
Tokyo, 2020.

Preface

Andy Liu and I have never met in person. We first corresponded in 2003 over a problem in combinatorial geometry. Since then we have exchanged e-mails about countless topics in recreational mathematics.

Over the years I have come to admire Andy for his industry, clear writing, and disciplined enthusiasm. Apart from his acclaimed work as a professor at the University of Alberta, Andy has traveled all over the world to encourage students and inspire them with love of mathematics. Sometimes he took the trouble to send me a picture postcard, which I was always pleased to get.

For years Andy has been sending me puzzles by Nob Yoshigahara. Sometimes Andy needed help with solving them, and I was happy to lend my programming skills to the task. I soon realized that Nob had been a genius at puzzles and had done much hard work to create them.

This book was Andy's idea, and I think it was a very good idea. Nob's puzzles are entertaining, challenging, and often amazing. They deserve to be better known. Andy's lucid solutions and commentary have enriched the book enormously. My own contributions were modest. This did not stop me from accepting Andy's kind offer to include me as an author.

I hope you will enjoy the book!

George Sicherman,
Wayside, New Jersey, 2020.

Preface

NOB is a household name and a legend in the worldwide community of mathematical and mechanical puzzles. Born on May 27, 1936, his full name was **Nobuyuki Yoshigahara**. He graduated from the Tokyo Institute of Technology in applied chemistry, and became a high school teacher in chemistry and mathematics. I could only envy his students who had such an interesting, ingenious and inspiring teacher.

I first met Nob in 1986 in Calgary at the 70th birthday party of Richard Guy, who has just passed away at the age of 103. It was a significant conference for me and I am sure for many other people. Many of my lifetime friendships began in the foothills of the Rocky Mountains. These are truly eminent people, and are all very humble and approachable.

One of them is **Kate Jones**, a Hungarian-American artist who designs and manufactures stunningly colorful and mesmerizing puzzles in wood and acrylic. They are available from her website *www.gamepuzzles.com*. Another is **Jerry Slocum**, puzzle historian and collector, and the founder of the International Puzzle Party. His private collection is the largest in the world, and it is going in installments to the Lilly Library of Indiana University, an invaluable donation and a true legacy. For further details, see *www.indiana.edu/∼liblilly*.

And of course, there was Nob! He was particularly outgoing. He always had a suitable puzzle to challenge the right person. His insightful mind saw things from many different angles, and he had an uncanny sense of humor which allowed him to present things in the most appealing manner. He invented numerous mechanical puzzles. Those best known in North America are *Rush Hour* and other products from Bill Ritchie's company *Binary Arts*, later renamed *Thinkfun*. See *www.thinkfun.com/*.

I next met Nob in 1991, in Los Angeles, at my first International Puzzle Party. It was an eye-opening experience. I saw for the first time a hint of the wide spectrum of Nob's genius. I bought many of his puzzles. I met Nob many times afterwards, either at the International Puzzle Parties or at the Gatherings for Gardner, another amazing event which was organized by the late Tom Rodgers.

I remembered vividly a talk which Nob gave at the 2000 International Congress of Mathematics Education in Tokyo. He was demonstrating how to construct a hollow die with transparent faces such that when it was rolled on an overhead projector, the spots on the top face would show but not those on the bottom face. He needed something like this for his classroom, and he just sat down and invented it.

In 2003, the Association of Game & Puzzle Collectors honored Nob with the Sam Loyd Award, for individuals who have made a significant contribution to the world of mechanical puzzles. On June 19, 2004, Nob left us. It was the passing of an era. In 2005, the puzzle design competition of the International Puzzle Parties was renamed the Nob Yoshigahara Puzzle Design Competition.

Nob had authored over eighty puzzle books. Only one of them, *Super-super Hard Puzzles*, had so far been translated into English. It was titled *Puzzle 101*, and was published by *A K Peters*, a progressive and innovative company. It was founded by the wife-husband team of Alice and Klaus Peters. Klaus was the first mathematics editor of Springer-Verlag, a publishing giant in Germany. Alice was their first North American mathematics editor.

I had met Klaus many times, both at the Gatherings for Gardner and at the Joint Mathematics Meetings of the Mathematical Association of America and the American Mathematical Society. We had become personal friends. The company became a sponsor of the Alberta High School Mathematics Competition, and the company logo lives on in the former sponsors' section on the website *www.ualberta.ca/mathematical-and-statistical-sciences/outreach/alberta -high-school-math-competition/past-logos.pdf*.

At one point, Klaus invited me to become a shareholder of *A K Peters*. He needed the capital to resist an attempted take-over. I was only too proud to have an opportunity to be involved. Then in 2010, he accepted an attractive offer from the British publishing company Taylor & Francis, after he was given assurance that the company could retain its autonomy. For the deal to go through, he had to regain all shares. Reluctantly, I relinquished mine, making a small profit.

In 2012, Alice and Klaus were abruptly dismissed and their company closed down. I am sure that the anguish led to Klaus's untimely passing on July 7, 2014. Alice had since found employment at the Museum of Mathematics in New York, with their outreach and publishing department.

From what I understand, the book *Puzzle 101* did not do well, and I can see why. The Japanese readers will be offended if full solutions are provided for the puzzles, while the North American audience was quite the opposite. Also, the problems came one after another without rhyme or reason. The book did not really do Nob justice.

I contacted **Takayuki Yoshigahara**, Nob's son, and obtained 200 problems from Nob's personal files. Most of them have appeared in his books in Japanese. Takayuki and I have organized them into ten chapters by topics, and provided full and motivated solutions to all the puzzles. I hope our effort would enhance Nob's well-deserved reputation as a Grand Master of puzzles.

We have had to struggle with many of the puzzles. I did not know that Nob was an avid computer programmer, but Takayuki informed me that he often used computers in solving mathematical puzzles. So I turned to another computer wizard for help. **Dr. George Sicherman** is fondly known as the Colonel, though he has assured me that he is neither in the military nor in the fried chicken business. A treasure trove for recreational mathematics is his website *userpages.monmouth.com/~colonel.*

We have relied on George for solving so many of the puzzles that this book is a true collaboration among the three of us. Since books written in parts by different people tend not to work well, I have done the actual writing under the guidance of Takayuki and George.

Our target audience includes bright young children. To ensure that at least the first three chapters are accessible to them, I sought advice from three youngsters. For Chapter One, my consultant was David Goh of Surabaya, then thirteen years old. For Chapter Two, my consultant was Nikolaj Hansen of Burnaby, also thirteen at the time. For Chapter Three, my consultant was Jacob Chu of Calgary, who was only eight years old then.

I thank Kate Jones for her proof-reading of a draft of the manuscript. Dr. Yoshiyuki Kotani has provided invaluable advice. He is the current President of the Japanese Academy of Recreational Mathematics, of which I am a proud member. He is also on the editorial board of the Academy's monthly Bulletin, another important source of wonderful ideas for puzzles.

I am grateful to Robinson dos Santos, Anne Comment, Jan Holland, Saveetha Balasundaram, Gomathi Mohanarangan, Jeffrey Taub and other staff members of Springer Nature for their encouragement, advice and support.

Andy Liu,
Edmonton, Alberta, 2020.

Comments

Andy Liu

The moving tribute in Takayuki's Preface to his father evokes memories of my own of this great man that I wish to share with everyone.

1. At the International Puzzle Party in 1993 in Breukelen, the Netherlands, Nob was one of the first three people inducted into the Puzzle Hall of Fame. He received as a gift a puzzle in the form of a metal bottle. He said to me, "Andy, I have to solve this puzzle before I go home, because the Japanese customs officers will ask me to open it. However, I can tell you something right now. There is no alcohol inside, or I would know."

2. A Danish friend of mine visited Nob in Tokyo shortly after his serious illness, and told me that she did not think he would be long for this world. Yet I met him at the International Puzzle Party in 1997 in San Francisco. I asked him how he managed to recover so well. He said, "Andy, I fired my doctor and went back to drinking again."

3. At the International Puzzle Party in 1991 in Los Angeles, I bought my favorite puzzle of Nob's. The container is a $26 \times 22 \times 17$ box. It is open at the top while the other five sides are of thickness 1. Six of the rectangular blocks are white. They are A($24 \times 10 \times 8$), B($24 \times 8 \times 5$), C($16 \times 12 \times 10$), D($16 \times 12 \times 5$), E($12 \times 10 \times 8$) and F($12 \times 8 \times 5$). The seventh or the extra green piece is the same size and shape as F. I was only able to fit the six white pieces into the box, as shown in the diagram below. There was simply no room for the green piece! A friend of mine thought that I was crazy to pay one hundred American dollars for this small wooden puzzle. My reply was that I was not buying lumber. What made it worth so much was the idea behind this masterpiece from Nob!

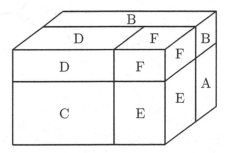

4. I sent Nob the following puzzle. Dissect the figure in the diagram below into three pieces and reassemble them into an equilateral triangle.

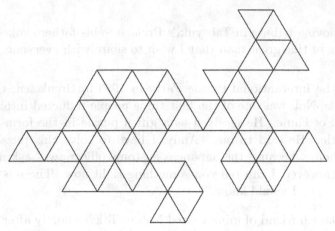

He wrote, "Several people in Japan have solved your puzzle, but Nob is not one of them." I wrote back, "How can that be? The idea of my puzzle is based on the idea of one of your puzzles."

My solution is shown in the diagram below. The interested readers may wish to find the original puzzle by Nob which is in this book.

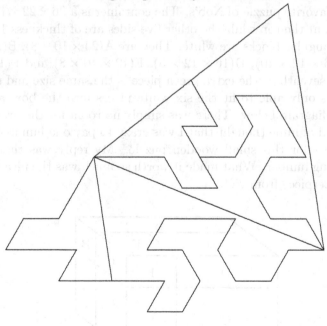

Table of Contents

Chapter One

Introductory Puzzles

Puzzle 1.
A girl put a bookmark in a novel she was reading. She first thought that she put it between pages 339 and 340. Then she thought that she put it between pages 388 and 389. Obviously, she could not be right both times, but she was not wrong both times either. Which time was she right?

Puzzle 2.
If it takes four cats to catch four mice in four minutes, how long would it take forty-eight cats to catch forty-eight mice?

Puzzle 3.
If you deposit $20 into the Giant Bank, you will get an interest of $2 in 2 years. If you deposit $30 into the Titan Bank, you will get an interest of $3 in 3 years. Which bank offers you the better deal?

Puzzle 4.
A rectangular block of chocolate consists of 24 pieces in a 4×6 array. In the first move, we may break up the block along a dividing line into two smaller rectangular blocks. In each subsequent move, we may break up any one of the blocks along a dividing line into two smaller rectangular blocks. What is the minimum number of moves before we have 24 blocks each consisting of a single piece of chocolate?

Puzzle 5.
Alice, Brian, Colin, Debra, Edwin and Fiona make the following statements:

Alice:	Only one of us is lying.
Brian:	Exactly two of us are lying.
Colin:	Exactly three of us are lying.
Debra:	Exactly four of us are lying.
Edwin:	Exactly five of us are lying.
Fiona:	All of us are lying.

Who are lying?

Puzzle 6.
Alice, Debra and Fiona are dating Brian, Colin and Edwin, not necessarily respectively. Alice does not date Brian, Debra does not date Colin, and Colin does not date Alice. What are the dating pairings?

© The Author(s), under exclusive license to Springer Nature Switzerland AG 2020
A. Liu et al., *The Puzzles of Nobuyuki Yoshigahara*, Problem Books in Mathematics,
https://doi.org/10.1007/978-3-030-62896-3_1

Puzzle 7.

Alice, Brian, Colin, Debra, Edwin and Fiona are sitting at a round table. Fiona is in the second seat to the left of Alice, and Debra is in the second seat to the left of Fiona. Brian is in a seat adjacent to Debra. Colin and Brian are in the two seats adjacent to Fiona. What is their seating arrangement relative to one another?

Puzzle 8.

Each of Alice, Brian, Colin, Debra and Edwin lives in a different one of five houses in a row. Edwin does not live next door to either Brian or Colin. Debra does not live next door to Colin and Edwin. Alice does not live next door to Brian and Colin. Who lives in the middle house?

Puzzle 9.

Tom is one of only five part-time workers in a warehouse, even though the following duty roster seems to suggest a larger workforce.

Day of the week:	Workers in charge:	Other workers:
Monday	Mr. Westwood	
Tuesday	Mr. Northcott	Dick
Wednesday	Mr. Eastwood	Bill and Fred
Thursday	Mr. Upton	Harry and Dick
Friday	Mr. Eastwood and Mr. Southcott	Harry and Dick
Saturday	Mr. Upton	Bill

A worker's last name is used when he is put in charge of the warehouse, but his first name is used otherwise. What is the full name of each worker?

Puzzle 10.

If the problem you solved before you solved the problem you solved after you solved the problem you solved before you solved this problem is more difficult than the problem you solved after you solved the problem you solved before you solved this problem, then is the problem you solved before you solved this problem more difficult than this problem?

Solutions

Puzzle 1.
She could not have been right the first time since pages 339 and 340 were printed back to back. Hence she must be right the second time.

Puzzle 2.
In four minutes, four cats will catch four mice. Hence one cat will catch one mouse, and forty-eight cats will catch forty-eight mice. So the time this takes is still four minutes.

Puzzle 3.
You earn 10% interest at the Giant Bank in 2 years and at the Titan Bank in 3 years. Obviously the Giant Bank offers the better deal.

Puzzle 4.
Each move increases the number of blocks by 1. Since we start with 1 block and end up with 24, the process always takes $24 - 1 = 23$ moves.

Puzzle 5.
Since the six statements contradict one another, at most one is true. Hence at least five of them are lying. However, they cannot all be lying, as otherwise Fiona will be telling the truth. Hence Edwin is the only one who is telling the truth, while the other five are lying.

Puzzle 6.
Since Colin does not date either Alice or Debra, he is dating Fiona. Since Alice does not date Brian, she is dating Edwin. Hence Brian and Debra are the third dating pair.

Puzzle 7.
Let us number the seats 1 to 6 in clockwise order and put Alice in seat 1. Then Fiona is in seat 3 and Debra in seat 5. Since Brian is in a seat adjacent to both Debra and Fiona, he is in seat 4. This puts Colin in seat 2 and leaves seat 6 for Edwin.

Puzzle 8.
Since Edwin does not live next door to Brian, Colin or Debra, he lives in an end house with Alice next door. Since Alice does not live next door to Brian and Colin, Debra must live in the middle house. The extra piece of information that Debra does not live next door to Colin puts Colin in the other end house, with Brian next door.

Puzzle 9.
We draw a chart of last names versus first names. The condition on Tuesday means that Dick's last name is not Northcott. So we put "Tu." in the slot of Dick Northcott to show that this is impossible. The completed chart is shown below.

3

Mr.	Bill	Dick	Fred	Harry	Tom
Eastwood	We.	Fr.	We.	Fr.	
Northcott		Tu.			
Southcott		Fr.		Fr.	
Westwood					
Upton	Sa.	Th.		Th.	

Since Mr. Eastwood is not Bill, Dick, Fred or Harry, he must be Tom. Then Mr. Upton must be Fred, Mr. Southcott must be Bill, Mr. Northcott must be Harry and Mr. Westwood must be Dick. We may also start with the deduction that Dick is Mr. Westwood, and continue from there.

Puzzle 10.

This is all about careful reading of very convoluted information. Let us call this problem B, and call A the problem you solved before you solved problem B. Then what is asked is whether problem A is more difficult than problem B. The given information tells us that a certain problem is more difficult than another problem. Let us find out what these two problems are. The former is "the problem you solved before you solved the problem you solved after you solved the problem you solved before you solved this problem". Let us simplify things in four steps to

1. "the problem you solved before you solved the problem you solved after you solved the problem you solved before you solved problem B";

2. "the problem you solved before you solved the problem you solved after you solved problem A";

3. "the problem you solved before you solved problem B";

4. "problem A".

Similarly, "the problem you solved after you solved the problem you solved before you solved this problem" simplifies to "problem B". So we are told that problem A is more difficult than problem B.

Chapter Two
Matchstick Puzzles

Puzzle 1.
Figure 2.1 shows 9 matchsticks forming an hourglass consisting of two equilateral triangles. Move 4 matchsticks to turn the hourglass upside down.

Puzzle 2.
Figure 2.2 shows 10 matchsticks forming an upside down chair. Move 2 matchsticks to turn the chair right side up.

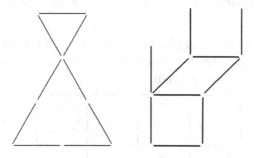

Figure 2.1   **Figure 2.2**

Puzzle 3.
Figure 2.3 shows 5 matchsticks forming a horse. Move 1 stick to make the horse face in a different direction.

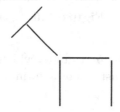

Figure 2.3

Puzzle 4.
Figure 2.4 shows 20 matchsticks forming 7 unit squares. Move 3 matchsticks so that only 5 unit squares are left. Each matchstick must be used to form one of the 5 unit squares.

Puzzle 5.
Figure 2.5 shows 24 matchsticks forming a 3 × 3 grid which contains many squares of various sizes. What is the minimum number of matchsticks that must be removed so that no squares are left?

5

© The Author(s), under exclusive license to Springer Nature Switzerland AG 2020
A. Liu et al., *The Puzzles of Nobuyuki Yoshigahara*, Problem Books in Mathematics,
https://doi.org/10.1007/978-3-030-62896-3_2

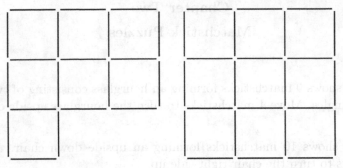

| Figure 2.4 | Figure 2.5 |

Puzzle 6.

Figure 2.6 shows 31 matchsticks forming a 3×4 grid which contains 20 overlapping squares of three different sizes. Remove as few matchsticks as possible so that no squares are left.

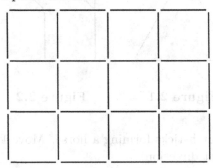

Figure 2.6

Puzzle 7.

Twelve matchsticks form a 3-4-5 triangle, which has area 6. Move five of them so that the twelve matchsticks now form a polygon which has area 2.

Puzzle 8.

(a) Twelve matchsticks form a 3-4-5 triangle. Use two matchsticks to bisect its area.

(b) Nine matchsticks form a 2-3-4 triangle. Use two matchsticks to bisect its area.

The matchsticks may not overlap or protrude outside the triangle.

Puzzle 9.

Arrange some matchsticks so that they only meet at the endpoints, and exactly three meet at every endpoint.

Puzzle 10.

Use nine matchsticks to form three unit squares.

Solutions

Puzzle 1.
Since we start with 2 horizontal matches on top and end up with 2 at the bottom, at least one horizontal match must be moved. There are two groups of 3 matches forming slanting sides. In each group, if 1 match is moved, all 3 must be moved. It follows that we must move at least 4 matches. Figure 2.7 shows how this can be done.

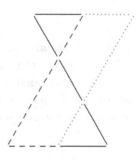

Figure 2.7

Puzzle 2.
Draw the chair right side up and compare with the upside down chair. The two matchsticks marked by dotted lines are to be moved to their new positions marked by dashed lines.

Figure 2.8

Puzzle 3.
In Figure 2.9 on the left, move the matchstick marked by a dotted line to its new position marked by a dashed line. The horse, resting on its hindquarters, is facing north. When it stands up, as shown in Figure 2.9 on the right, it is facing east instead of facing west.

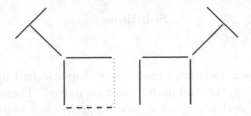

Figure 2.9

Puzzle 4.

Since we want to be left with five unit squares formed with 20 matchsticks, each is made with 4 matchsticks. It follows that no two of the unit squares can share a common side. There are four such squares in the original diagram. To minimize the number of matchsticks being moved, we should keep all of them. The fifth square must be formed where the loose edge is, as shown in Figure 2.10.

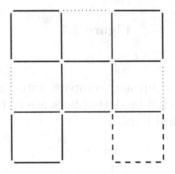

Figure 2.10

Puzzle 5.

We first prove that the removal of five matchsticks is insufficient. We must remove at least one matchstick forming the big 3×3 square, and at least one forming the central unit square. Consider the other eight unit squares. At most two have been destroyed so far. If we are to remove only three more matchsticks, each must destroy two of the other six, so that they form three adjacent pairs. In Figure 2.11, the two unit squares already destroyed are shaded. The matchsticks removed to destroy them are replaced by dashed lines, with a choice at the bottom left corner. The situation where the shaded squares are adjacent is shown in Figure 2.11 on the left, while the situation where they are not is shown in Figure 2.11 on the right. The matchstick which divides the two squares in each adjacent pair is removed, replaced by a dotted line. In either case, the 2×2 square at the upper left corner remains intact.

8

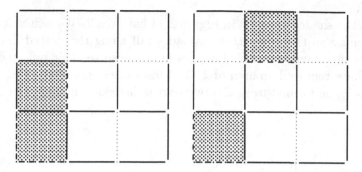

Figure 2.11

Figure 2.12 shows how the removal of six matchsticks can destroy all squares.

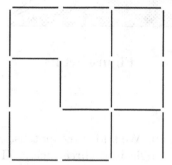

Figure 2.12

Puzzle 6.
From each 1×1 square, we must remove 1 matchstick. Since there are 12 of them, and each matchstick belongs to at most 2 of them, we must remove at least 6 matchsticks. If we remove only 6 matchsticks, we must partition the grid into 6 dominoes and remove the central matchstick from each. However, 2 of the dominoes will form a 2×2 square. Hence we must remove at least 7 matches. Figure 2.13 shows a way to do so.

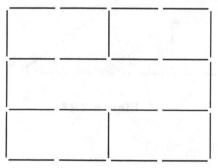

Figure 2.13

Puzzle 7.

Note that the shaded triangle in Figure 2.14 has area 2. We reflect it across its hypotenuse so that the other two sides fall along the dotted lines. By moving the five matchsticks below this hypotenuse to their reflected positions, we have removed an area of 2+2=4 from the original triangle, so that the new polygon formed from the twelve matchsticks will have area 2.

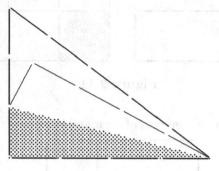

Figure 2.14

Puzzle 8.

(a) The triangle has area 6. We add three matchsticks, one along a dashed line, dividing the triangle into three pieces. The square has area 1. Since the darkly shaded triangle has area 1, the piece above the square has area 2. Since the lightly shaded triangle has area $1\frac{1}{2}$, the piece to the right of the square has area 3. If we now remove the matchstick along the dashed line, the remaining two matchsticks will bisect the area of the triangle.

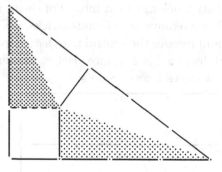

Figure 2.15

(b) The base of the top triangle in Figure 2.16 joins the midpoints of two of the edges of the whole triangle. Hence its length is exactly two matchsticks and its area is exactly one quarter the area of the whole triangle. The shaded parallelogram below it has the same area since it has the same height but only half the base. It follows that the two additional matchsticks shown by solid lines bisect the area of the triangle.

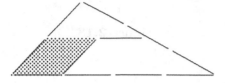

Figure 2.16

There is another solution. The shaded parallelogram in Figure 2.17 has one quarter the area of the whole triangle. The triangle above it has the same area since it has the same height but double the base. The dotted lines, which form a mirror image of the whole triangle, show that the third side of this triangle has the length of exactly one matchstick. It follows that the two additional matchsticks shown by solid lines bisect the area of the triangle.

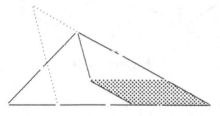

Figure 2.17

Puzzle 9.
Venturing into three-dimensional space, we can use six matchsticks to form the sides of a tetrahedron. The condition of the puzzle is then met. Staying in the plane, five matchsticks form two unit equilateral triangles with a common side. At two of the endpoints, we do have three matchsticks meeting there. At the other two endpoints, we only have two matchsticks meeting there. So we draw two such configurations, and use two matchsticks to connect one deficient endpoint on each side, as shown in Figure 2.18.

11

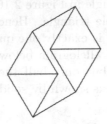

Figure 2.18

Puzzle 10.

Venturing into three-dimensional space, a triangular prism can be formed from nine matchsticks, as shown in Figure 2.19 on the left. It has three lateral faces which are squares. Another solution consists of three faces of a cube sharing a common vertex, marked by the black dot in Figure 2.19 on the right.

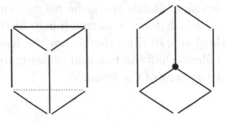

Figure 2.19

Staying in the plane, three unit squares standing alone require 12 matchsticks. We may save 1 matchstick if two squares share a side. However, two squares can share at most one side. To save 3 matchsticks, all three pairs of squares must share sides. This is not possible. However, if we allow overlapping squares, there is a solution as shown in Figure 2.20.

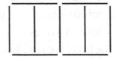

Figure 2.20

Chapter Three
Maze Puzzles

Puzzle 1.

Figure 3.1 shows the floor plan of the Art Museum. It was robbed overnight, and all paintings from the three galleries were removed. The security system was on, and an alarm would have been triggered had any section of the corridors been traversed more than once. Somehow, the robbers got around that problem. What was their route if they had to come in through the entrance and escape through the exit?

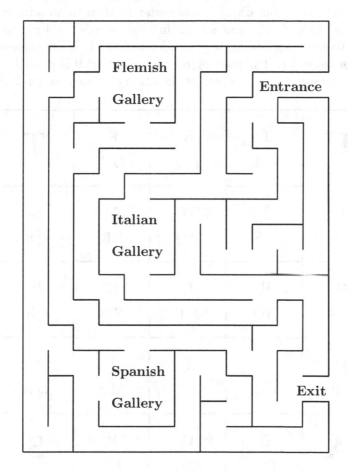

Figure 3.1

13

Puzzle 2.

A library in Graz has twenty-five rooms in a five-by-five configuration. All of the books under A are stored in the room in the southwest corner. All of the books under U are stored in the room in the northwest corner. Most of the books under S are stored in the room in the southeast corner. Most of the books under T are stored in the room in the northeast corner. The books under B to R, as well as some books under S and T, are scattered in the remaining rooms, each of which contains books under exactly three different letters. The books under V to Z have been moved to Vienna. The main entrance to the library is at the southwest corner containing the books under A. The librarian makes the rule that from there, readers can move only to an adjacent room containing books under B, then to an adjacent room containing books under C, and so on. In other words, readers must move from room to room containing books under letters which are consecutive in the alphabetical order. The readers soon discover that it is possible to reach only one of the other three corners of the library. Which corner is it?

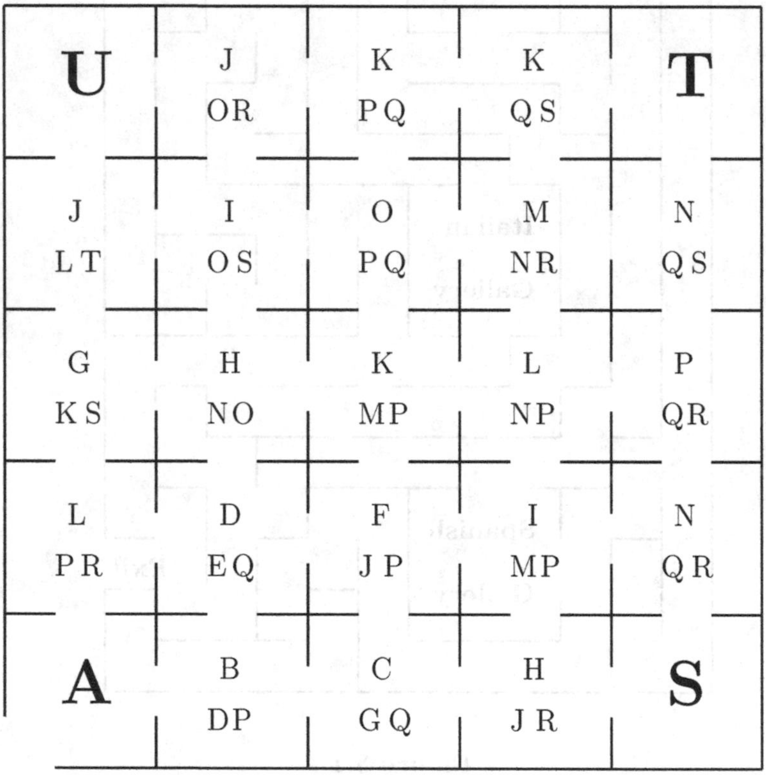

U	J OR	K PQ	K QS	T
J LT	I OS	O PQ	M NR	N QS
G KS	H NO	K MP	L NP	P QR
L PR	D EQ	F JP	I MP	N QR
A	B DP	C GQ	H JR	S

Figure 3.2

Puzzle 3.

Figure 3.3 shows the ground of a Six-Nation Trade Fair, hosted by Japan and with participation from India, Korea, Laos, Mongolia and Nepal. Paths connect seventeen tents labeled by the initial letters of the countries putting up the respective tents. Entrance to the Fair is via one of the three northern tents, and exit is via one of three southern tents. Visitors can only enter a northern tent from outside, and must exit from a southern tent. A ticket allows admission to six tents, and a tent may not be passed through without admission. How can one visit one tent put up by each country?

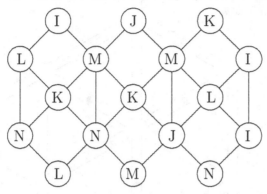

Figure 3.3

Puzzle 4.

Starting from the white circle in Figure 3.4, move along the paths until the black circle is reached. If no path may be traversed more than once, what is the maximum length, in terms of unit blocks, of the whole tour?

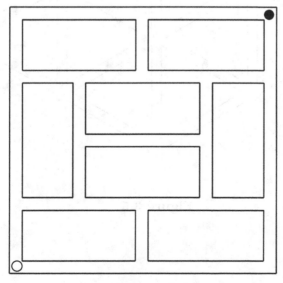

Figure 3.4

15

Puzzle 5.

In the Garden of Eden, which is shaped like a hexagon, there are twelve apples, as indicated by the black circles in Figure 3.5. You may enter the Garden of Eden from either end and exit via the other end, and move along the garden paths to collect as many apples as you can. However, if you ever visit the same intersection of the garden paths more than once, you will be struck down by a thunderbolt. What is the maximum number of apples you can collect?

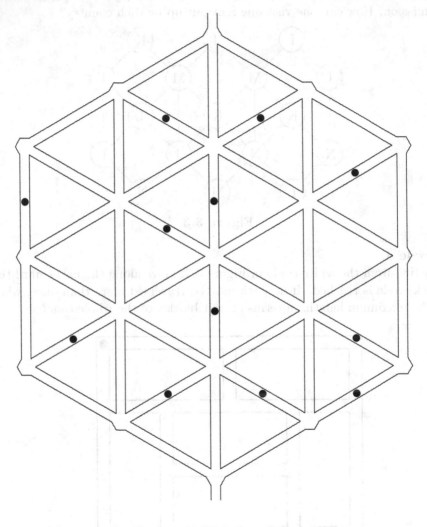

Figure 3.5

Puzzle 6.

Figure 3.6 shows the map of Windsor Castle. The twenty circles represent towers. Various towers are connected by walls. The entrance from outside is through the outer north tower, as indicated by an in-arrow. The entrance to the central courtyard is through the inner south tower, as indicated by an out-arrow. The number inside each circle represents the admission price, in pounds sterling, for that tower. Visitors may not pass through a room without paying. A staircase in the central courtyard descends to the dungeon where the Crown Jewels are exhibited. Mary is not interested in the exhibits in the towers. She only wants to find the cheapest path to the central courtyard. How can she reach the central courtyard as cheaply as possible?

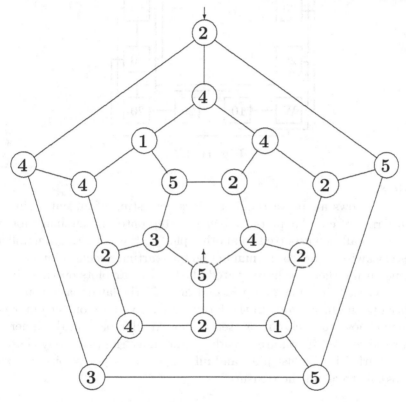

Figure 3.6

Puzzle 7.

The Crown Jewels are exhibited in the dungeon of Windsor Castle, the floor plan of which is shown in Figure 3.7. The paths do not intersect. They go across one another by over- and underpasses. A staircase descends from the central courtyard into the welcome room marked W, and a tunnel from the exit room marked E is the only way for visitors to leave.

17

Each of the ten exhibit rooms has an admission charge of a number of pounds sterling marked on the floor plan. Visitors may not pass through a room without paying. In the welcome room, Mary discovers that she has lost her purse, and has much less money in her possession than she had thought. Thus she has to find a way to get to the exit by spending as little money as possible. How can she reach the exit as cheaply as possible?

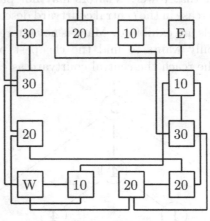

Figure 3.7

Puzzle 8.
Figure 3.8 shows an Incan religious grove consisting of an outer ring and an inner ring of circular plots. When you step onto any circular plot, you will be magically transported to another plot on the same ring a number of spaces clockwise, equal to the number on the starting circular plot. This will continue until either you have visited all the circular plots on that ring, or you have visited the same plot a second time. In the latter case, you will be sacrificed to the Incan god. In the former case, if you are on the outer ring, you may choose to back out or step onto a circular plot on the inner ring of your choice, and if you are already on the inner ring, you may claim the treasure buried in the last plot, and all magical transportations will cease. Is it possible to claim the treasure?

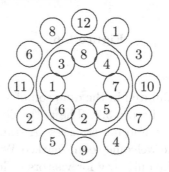

Figure 3.8

Puzzle 9.

Roger, Yvonne, Brian, Gloria and Wally are in an underground catacomb which has 64 rooms in an 8 × 8 configuration. Each starts from either of the two rooms labeled with their initial, as shown in Figure 3.9. They must move from room to an adjacent room in the same row or column to reach the other room with her or his initial. However, once one of them visits a room, it becomes inaccessible. How can they accomplish the task?

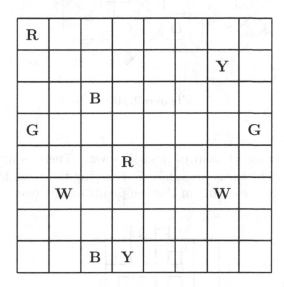

Figure 3.9

Puzzle 10.

Figure 3.10 shows fourteen rectangular pieces of paper lying on a flat surface and overlapping one another. Beginning on the piece marked B, move from piece to adjacent piece in order to finish at the piece marked F. The path must alternately climb up to a piece of paper stacked higher and come down to a piece of paper stacked lower. The same piece may be visited more than once, and it is not necessary to visit every piece. List the pieces of paper in the order visited.

19

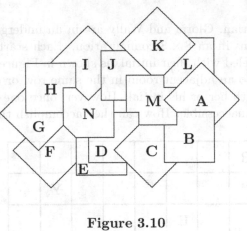

Figure 3.10

Puzzle 11.

Figure 3.11 is the street map of a small town. There is a very strange traffic rule. No right turns are allowed. Entering the town from point E, it is possible to exit from any of the five points except one. Which exit is impossible?

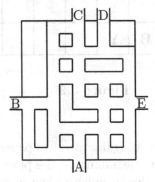

Figure 3.11

Puzzle 12.

Figure 3.12 is the street map of a small town. There is a very strange traffic rule. No turns are allowed at any intersection unless it is impossible to drive straight on. Then both left turns and right turns, if available, are allowed. Entering the town from point E, it is possible to exit from any other point except one. Which exit is impossible?

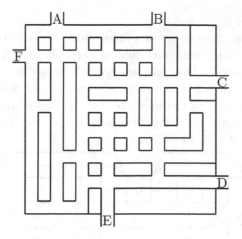

Figure 3.12

Puzzle 13.

Figure 3.13 is the street map of a small town. There is a very strange traffic rule. No turns are allowed at any intersection unless it is impossible to drive straight on. Then both left turns and right turns, if available, are allowed. Entering the town from point E, show how to exit from point X.

Figure 3.13

21

Puzzle 14.

Figure 3.14 shows the map of a town consisting of blocks connected by 31 streets, with shops along only one side of each street. A girl who loves shopping lives in Block G. One day, she intends to go along all the streets in the town exactly once, which she is able to do, and asks her boyfriend to pick her up at the end of the day. He does not know in what order she will be visiting the shops, but nevertheless he figures out at which block he should wait for her. Which block is it?

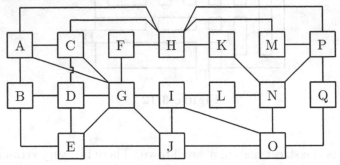

Figure 3.14

Puzzle 15.

Figure 3.15 shows the map of a county consisting of 23 towns connected by roads, with a video arcade in each town. A boy who loves video games works in an office in the town O. Every day after work, he visits all the towns exactly once before arriving at his home in the town H, spending much time at the video arcades. This means that he is neglecting his girlfriend. One day, she decides to meet him on his way home, but he varies his routes all the time. However, she has discovered that there is one road which he must use, so that she can wait for him along it. Which road is that?

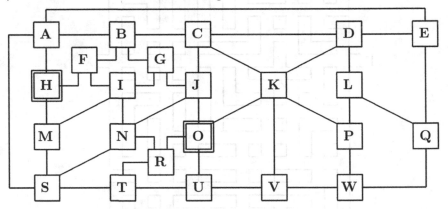

Figure 3.15

Puzzle 1.

Let us first shade all the dead-ends. The possible choices of paths are now much clearer and the route for the robbers emerges, as shown in Figure 3.16.

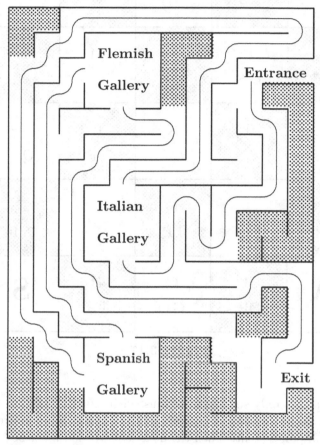

Figure 3.16

Puzzle 2.

First, we shade the rooms in the chessboard pattern, as shown in Figure 3.17. Then the letters in odd positions in the alphabet must be in unshaded rooms if they are to lie on the library tour. Similarly, the letters in even positions must be in shaded rooms. This allows us to eliminate the letters that cannot lie on the library tour. In particular, the northeast corner is inaccessible. The initial segment of the library tour, indicated by letters in bold face, is uniquely determined until we reach the letter L. Now there are three choices of the direction for continuation, but two of them lead to dead-ends quickly.

The remaining part of the library tour is also unique, as indicated by the letters in italics. Thus only the northwest corner can be reached.

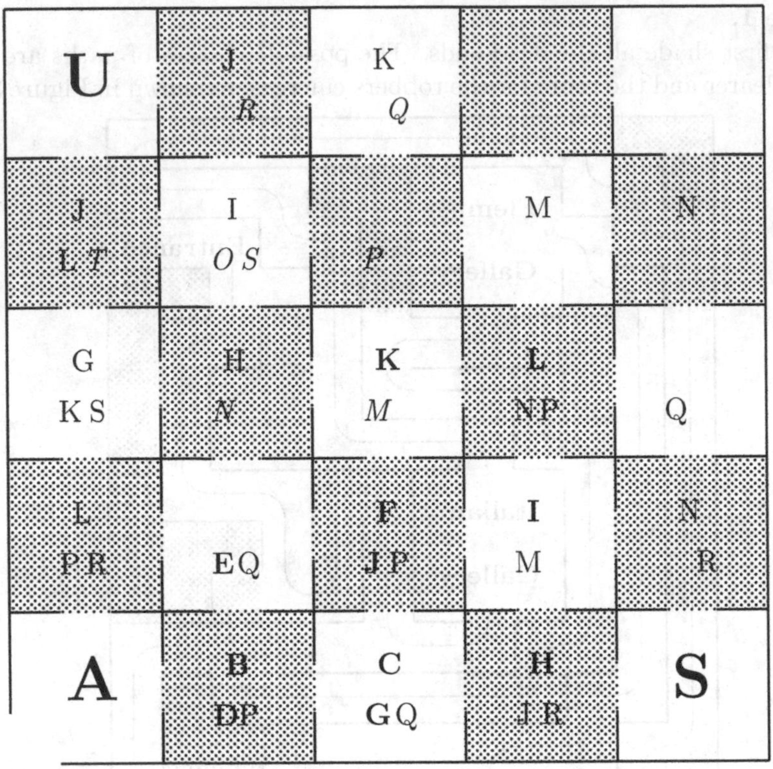

Figure 3.17

Puzzle 3.

With admission to only six tents, we must visit exactly one tent put up by each country. Note that Japan puts up only two tents. Suppose we enter the ground via a Japanese tent. Then the next tent we visit must be Mongolian. If we go left, or go right and then left, we will not get to visit an Indian tent. If we go right and then right again, we will not get to visit a Korean tent. It follows that we do not enter via a Japanese tent, and must therefore visit the other one. Suppose we enter via an Indian tent. If we go left, we must visit the Korean tent south of it, and cannot get to the Japanese tent. If we go right and then down, or go right twice and then left, we must exit via a Laotian tent and again cannot get to the Japanese tent. So we must go right three times, but now we exit either too early or too late. It follows that we must enter via a Korean tent, and are confined to the seven tents on the right. There are two ways to accomplish the task. We can visit in the following order: Korean, Indian, Laotian, Mongolian, Japanese and Nepalese, or in the alternative order: Korean, Mongolian, Japanese, Laotian, Indian and Nepalese.

24

Puzzle 4.

There are 32 blocks altogether. Figure 3.18 shows a tour in dotted lines consisting of 24 blocks. We now prove that this is indeed maximum. We must miss one of the two blocks at each circle. Clearly the shorter one is to be missed, and we shade them in the diagram below. There are also six intersections, marked by black dots, where three paths come together. One of the three must be missed. Since we have twelve black dots and each missed path accounts for two of them, we must miss an additional 6 blocks of paths.

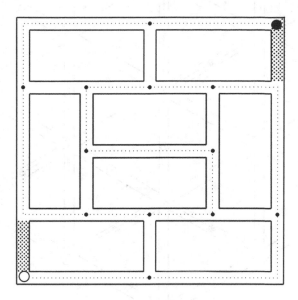

Figure 3.18

Puzzle 5.

If we go from one end straight to the other end, we will pass through three six-way intersections. For each of them, three of the garden paths meeting there contain an apple. Since you may visit this intersection only once, we cannot collect all three. It would appear that we have to omit three apples over all. However, the apple marked by a times sign in the diagram below lies on a garden path joining two of these intersections. If we omit it, it is necessary to omit only one of the two apples marked by plus signs. For other intersections where two garden paths containing apples meet, we must use both of them. It is easy to see how these pieces must be connected to give us the unique way of collecting ten apples.

25

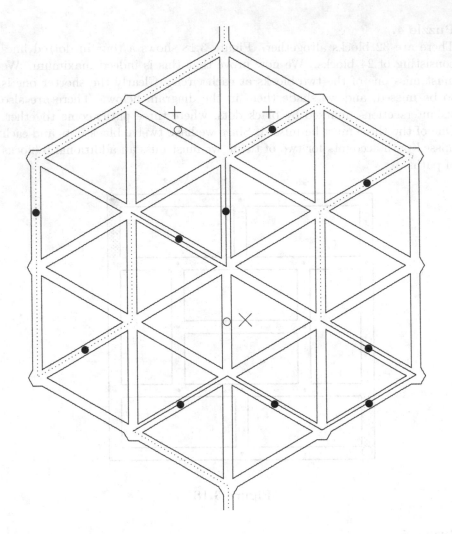

Figure 3.19

Puzzle 6.

We first wipe out all the numbers, leaving only the 5 in the inner south tower. This is because we must pay 5 pounds to enter it. There are three approaches to this tower. The cheapest one is coming from the south, at a cost of 2 pounds. This brings the total to 7 pounds, and we label this tower 7. Now there are four approaches to the two towers already labeled. This time, there are two cheapest paths at a cost of 5+3=7+1 pounds. We label those two towers 8. Continuing this way, we backtrack to the outer north tower. The final total is 5+2+1+2+2+5+2=19, and the cheapest path is marked by numbers in boldface in Figure 3.20.

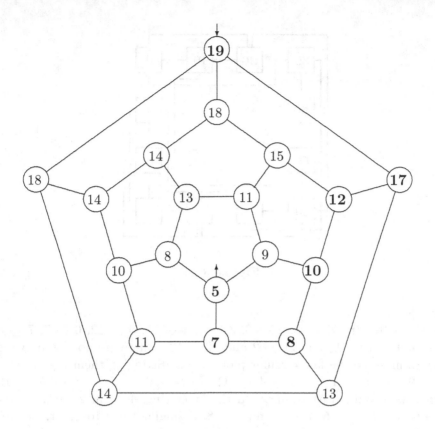

Figure 3.20

Puzzle 7.
We first wipe out all the numbers. We label the exit room 0 since there is no admission charge for entering it. There are three ways to get to this room. The cheapest one is coming from the west, at a cost of 10 pounds. We label this room 10. Continuing this way, we backtrack to the welcome room which is labeled 50. The unlabeled rooms all have total costs of at least 60 pounds to reach the exit room. From the welcome room, the cheapest path is marked by numbers in boldface in the diagram below, for a total cost of 30+20=50 pounds.

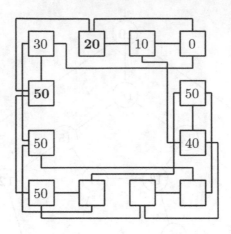

Figure 3.21

Puzzle 8.

On the outer ring, $1 \rightarrow 3$, $2 \rightarrow 6$, $3 \rightarrow 4$, $4 \rightarrow 11$, $5 \rightarrow 12$, $6 \rightarrow 7$, $7 \rightarrow 8$, $8 \rightarrow 5$, $9 \rightarrow 10$, $10 \rightarrow 1$, $11 \rightarrow 2$ and $12 \rightarrow 12$. There are no closed cycles, and 12 must be the last circular plot visited. Stringing them together, we have $9 \rightarrow 10 \rightarrow 1 \rightarrow 3 \rightarrow 4 \rightarrow 11 \rightarrow 2 \rightarrow 6 \rightarrow 7 \rightarrow 8 \rightarrow 5 \rightarrow 12$. Hence we should start from 9. In the inner ring, $1 \rightarrow 3$, $2 \rightarrow 1$, $3 \rightarrow 7$, $4 \rightarrow 6$, $5 \rightarrow 8$, $6 \rightarrow 5$, $7 \rightarrow 4$ and $8 \rightarrow 8$. Stringing them together, we have $2 \rightarrow 1 \rightarrow 3 \rightarrow 7 \rightarrow 4 \rightarrow 6 \rightarrow 5 \rightarrow 8$. Hence we should move onto 2 from the outer ring. We may then claim the buried treasure.

Puzzle 9.

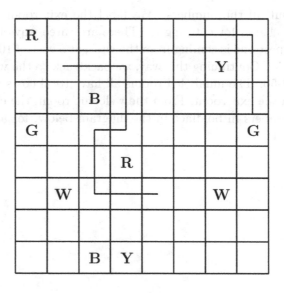

Figure 3.22

28

Suppose Gloria's path runs along the fourth row. Then each of Roger's, Brian's and Yvonne's paths will be cut off. So Gloria's path must rise above Brian's and Yvonne's paths and dip below Roger's path. Thus parts of her path must be as shown in Figure 3.22. It is now clear how the five paths must run, as shown in Figure 3.23.

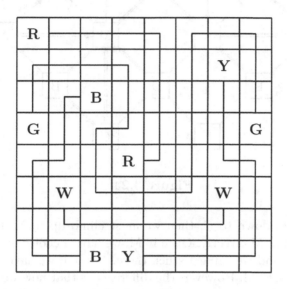

Figure 3.23

Puzzle 10.
We first reorganize the data, separating the pieces of paper. We draw an arrow from one piece of paper to another if we can come from the first down to the second. Note that each of M and N is connected to 7 other pieces, each of D and J is connected to 4 other pieces, while each of the others is connected to 3 other pieces. The path we seek consists of alternately going along with the arrow and going against it. Of the three arrows at A, the one from B cannot be used as otherwise we would be stuck at A. Equally useless are the arrows from L to M, from M to K, from I to N, from N to H, from G to F, from E to F, and from C to D. In Figure 3.24, they are drawn as single arrows while the usable ones are drawn as double arrows.

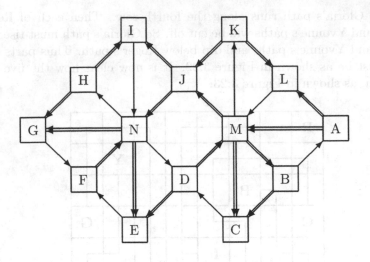

Figure 3.24

From B, if we go down to C, then we must climb up to M, but we could have climbed up to M directly. Once in M, we have a choice of C or J, but C leads back to B, and we will get stuck there. From J, we have to move onto K, L, A and back to M. However, the difference is that now we are climbing up, to D. The path continues with E, N, G, H, I, J, N and F. Thus the path is BMJKLAMDENGHIJNF.

Puzzle 11.
Some streets are accessible in only one direction, and others not at all. The former are indicated by arrows while the latter are shaded, as shown in Figure 3.25. This makes it clear that exit from point B is impossible.

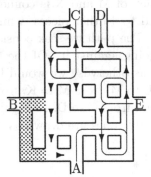

Figure 3.25

Puzzle 12.

The same approach used in Puzzle 11 reveals that all but point C are possible exits, as shown in Figure 3.26.

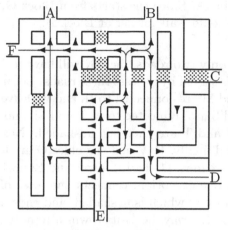

Figure 3.26

Puzzle 13.

The same approach used in Puzzle 11 yields the route shown in Figure 3.27.

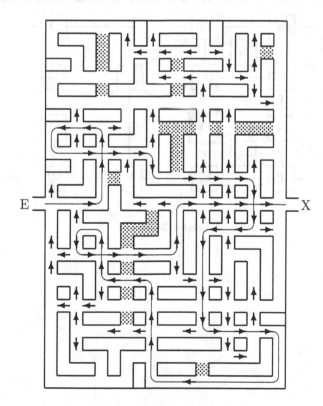

Figure 3.27

Puzzle 14.

Every block except Block B and Block G is connected to an even number of other blocks. The girl does not start from such a block, and therefore cannot end in such a block. Since she starts from Block G, she must end in Block B. So the boy should wait for her at Block B.

Puzzle 15.

If a town is served by only two roads, the boy will have to use both of them. However, every town is served by at least three roads. Let us shade the home H. The towns A, F and M are connected to H, and we leave them unshaded. The towns B, E, I and S are connected to A, F and M, and are shaded. The towns C, D, G, N, Q and T are then left unshaded. Next, J, K, L, R, U and W are shaded, and finally, P, V and the office O are left unshaded. As shown in Figure 3.28, we have 11 shaded towns and 12 unshaded ones. Note that every road, with one exception, connects one town of each type. The boy starts from the office O which is unshaded, and ends up in the home H which is shaded. Along the way, he visits towns alternating in type, except that he must go from an unshaded town to another unshaded town once. This is possible only between towns C and D. So the girlfriend should wait along the road connecting them.

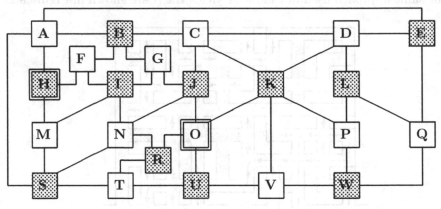

Figure 3.28

Chapter Four
Algorithmic Puzzles

Puzzle 1.
The queen, the prince and the princess were kidnapped by an evil wizard and locked up in a tower on an island. The stairwell was destroyed, and food was brought up to them by a pulley with two baskets. If the weight in one basket exceeded the weight of the other basket by more than 9 kilograms, the heavier basket would crash to the ground and kill anyone in it. The queen weighed 63 kilograms, the prince 117 kilograms and the princess 54 kilograms. The prince found and repaired a robot which weighed 45 kilograms, and it could withstand the crash. The robot worked out a way for all of them to get down to the ground. How could this be done?

Puzzle 2.
To get off the island, the queen, the prince, the princess and the robot must cross a narrow bridge which had just enough room for two of them to be on it at the same time. It was 18 minutes to midnight, and visibility was nil. Fortunately the robot had a lantern. Someone crossing the bridge must be holding it, and it must be brought back if there were still people on the island. The robot, the prince, the princess and the queen could cross the bridge in 1, 2, 5 and 10 minutes respectively. When two people crossed together, they moved at the speed of the slower one. The evil wizard would wake up at midnight. Could they get off the island before he could intervene?

Puzzle 3.
The capacities of jugs A, B, C and D are 12, 7, 5 and 3 cups respectively. Initially, jug A is full of wine but the other three are empty. The jugs have no internal markings. When wine is poured from one jug to another, exact measurement is only possible if we pour until either the first jug is empty or the second jug is full, whichever happens first. How can the wine be distributed among jugs A, B and C with 4 cups in each?

Puzzle 4.
Figure 4.1 shows a portion of a rail yard. An engine E is in square 4. Two carriages A and B are in squares 2 and 6 respectively. Only E has locomotive power, and it can move in either direction. It can also be coupled with either A or B, or both, in any combination, and push or pull them along. There is a narrow pass in square 11. A and B can go through it, but E may not enter square 6. A coupled car can turn corners except going from squares 4 and 3 to squares 3 and 8, and vice versa. Similarly, a coupled car may not go from squares 5 and 4 to squares 9 and 5, and vice versa. Move E, A and B so that A is in square 6, B is in square 2 and E is back in square 4.

© The Author(s), under exclusive license to Springer Nature Switzerland AG 2020
A. Liu et al., *The Puzzles of Nobuyuki Yoshigahara*, Problem Books in Mathematics,
https://doi.org/10.1007/978-3-030-62896-3_4

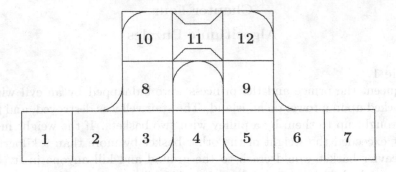

Figure 4.1

Puzzle 5.

A small zoo in Africa has only three cages, labeled N, O and B respectively. Each opens into the courtyard, and there is also a door between the O-cage and each of the other two cages, as shown in Figure 4.2.

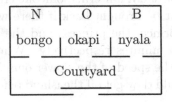

Figure 4.2

Three animals are acquired, a nyala, an okapi and a bongo. They all look alike, and the zookeeper put them arbitrarily in the cages. It is pointed out later that while the Okapi is indeed in the O-cage, the one in the N-cage is actually the bongo while the nyala is in the B-cage, as shown in Figure 4.2. The zookeeper decides to put things right by moving one animal at a time between two spaces connected to each other. However, he cannot manage two animals in the same space at the same time. What is the minimum number of moves to have all three animals in the right cages?

Puzzle 6.

Later, a yeti is imported from Asia. The courtyard is widened, and the middle part is carved out to make a Y-cage. The new-look zoo is shown in Figure 4.3, with two four-way doors.

34

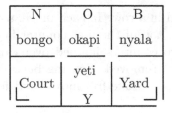

N	O	B
bongo	okapi	nyala
Court	yeti Y	Yard

Figure 4.3

When the animals are put back into the cages, the same mistake is made once again. As before, the animals can be moved one at a time between two spaces connected to each other, with no two in the same space at the same time. What is the minimum number of moves to have the nyala and the bongo trade places, leaving the okapi and the yeti in their correct cages?

Puzzle 7.
Arrange eight counters to form the configuration in Figure 4.4 on the left. Move the counters one at a time to form the configuration in Figure 4.4 on the right. A counter can only be moved if in its new position it is touching at least two other counters, and if it touches exactly two other counters, then they must not be in a line with it. A counter may not be lifted and then dropped from above.

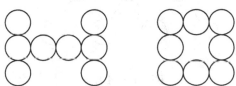

Figure 4.4

Puzzle 8.
Arrange eight counters to form the configuration in Figure 4.5 on the left. Move the counters one at a time to form the configuration in Figure 4.5 on the right. A counter can only be moved if in its new position it is touching at least two other counters, and if it touches exactly two other counters, then they must not be in a line with it. A counter may not be lifted and then dropped from above.

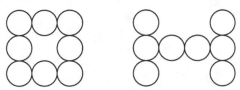

Figure 4.5

35

Puzzle 9.

Ten counters are arranged in a row. In each move, a counter can jump over two counters and land on top of the first counter beyond, forming a stack of two counters. The two counters jumped over can be separate or form a single stack, and are not removed. Spacing between counters and stacks has no effect on the jumps. Make five moves and form five stacks of two counters.

Puzzle 10.

Place one counter in each circle except one in Figure 4.6. Each counter may jump over an adjacent counter provided that the circle immediately beyond is vacant. The jumping counter will land on that circle while the jumped counter is removed. The jump may be in any of the three directions indicated by the lines. Remove all but one counter. Use as few jumps as possible, where a sequence of consecutive jumps by the same counter counts as a single jump.

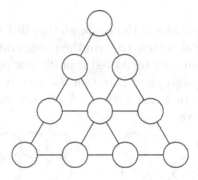

Figure 4.6

Puzzle 11.

Place one black counter and eight white counters on a 5 × 5 board as shown in Figure 4.7. Each counter may jump over an adjacent counter provided that the square immediately beyond is vacant. The jumping counter will land on that square while the jumped counter is removed. The jump may be horizontal, vertical or diagonal. Remove all eight white counters this way, leaving the black counter in its initial position. Use as few jumps as possible, where a sequence of consecutive jumps by the same counter counts as a single jump.

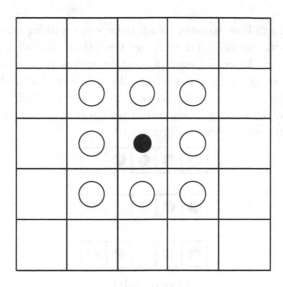

Figure 4.8

Puzzle 12.
Three black counters and three white counters are placed on a board consisting of seven squares, as shown in Figure 4.9. In each move, a counter can go to the vacant square either by moving there directly from an adjacent square in the same row or the same column, or by jumping over an intervening square which is necessarily occupied. The counter jumped over is not removed. Make the black counters and the white counters trade places in as few moves as possible.

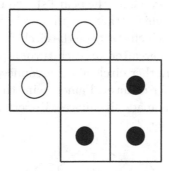

Figure 4.9

37

Puzzle 13.

On a 1×5 board are four counters, each occupying a different square. Each counter is black on one side and white on the other. Initially, they all have their black side up. In each move, a counter moves to the vacant square. All counters it passes along the way are flipped over, but not the moving counter itself. The objective is to have all counters with their white side up. They do not have to occupy their initial positions. The starting positions for three problems are shown in Figure 4.10.

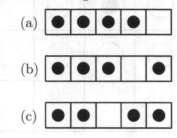

Figure 4.10

Puzzle 14.

Pamela and Robert play a tic-tac-toe game where each player has three counters of one color. They take turns placing one of their counters on the board, Pamela going first. A player wins if his or her three counters form a row, a column or a diagonal. If neither player has won by the time all counters have been placed, they continue the game by taking turns moving their own counters, again Pamela going first. A counter may only be moved from one square to an adjacent square on the same row or the same column, but not the same diagonal. Prove that Pamela can force a win.

Puzzle 15.

Pamela and Robert play a game on the board shown in Figure 4.11. Pamela controls a Police counter which starts on the circle marked P, and Robert controls a Robber counter which starts on the circle marked R. Robert moves first, and turns alternate thereafter. In each turn, a player moves his or her counter from a circle to another circle along a connecting line. If the Police counter lands on the Robber counter, Pamela wins the game. If this has not happened after Pamela has made five moves, Robert wins the game. Which player has a winning strategy?

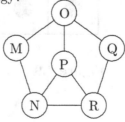

Figure 4.11

Solutions

Puzzle 1.

The task could be accomplished in eleven steps.

1. The robot crashed to the ground.

2. The princess came down while bringing the robot up. The difference in weight was $54 - 45 = 9$ kilograms, and the princess was safe.

3. The queen came down while bringing the princess up. The difference in weight was $63 - 54 = 9$ kilograms, and the queen was safe.

4. The robot crashed to the ground.

5. The prince came down while bringing the queen and the robot up. The difference in weight was $117 - 63 - 45 = 9$ kilograms, and the prince was safe.

6. The robot crashed to the ground.

7. The princess came down while bringing the robot up.

8. The queen came down while bringing the princess up.

9. The robot crashed to the ground.

10. The princess came down while bringing the robot up.

11. The robot crashed to the ground.

Puzzle 2.

There were five crossings altogether, three towards shore with two people each time, and two back to the island with one person each time. If the two people crossing alone were the same one, then it had taken part in all five crossings. It might as well be the robot. The total time would be $1+1+2+5+10=19$ minutes, which was too much. Suppose these two people were different. Then each had taken part in three crossings. They might as well be the robot and the prince. The minimum time for the three crossings with two people would be $2+2+10=14$ minutes, and the minimum total time would be $1+2+14=17$ minutes. This might be achieved by having the robot and the prince go first and the robot come back. Then the princess and the queen went together, and the prince came back. The final crossing was the same as the first one.

Puzzle 3.
We can leave 4 cups of wine in jug A by pouring wine into jugs C and D. We now pour wine into jug B, first from jug D and then from jug C, filling jug B and leaving 1 cup of wine in jug C. Next, we fill jug D from jug B, leaving the desired amount in the latter. Finally, the 3 cups in jug D are poured into jug C.

Puzzle 4.
First, E moves to 5, couples with B, moves to 3, couples with A, and then moves to 2, arriving at the position shown in Figure 4.12.

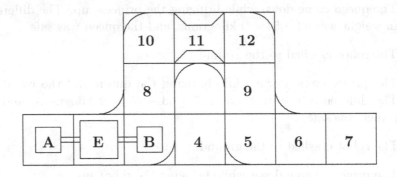

Figure 4.12

Next, E decouples from A and moves to 10, arriving at the position shown in Figure 4.13.

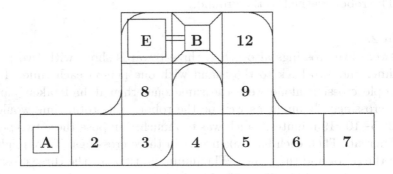

Figure 4.13

Then E decouples from B, moves back to 2 and couples with A again, moves to 6 in order to back up to 9, and has A couple with B, arriving at the position shown in Figure 4.14.

40

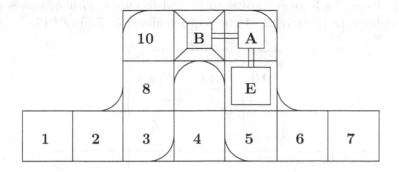

Figure 4.14

Now E moves to 7 in order to back up to 4, arriving at the position shown in Figure 4.15.

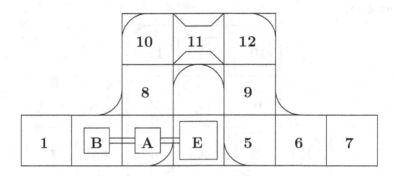

Figure 4.15

Next E has A decouple from B, moves to 6 in order to back up to 12, arriving at the position shown in Figure 4.16.

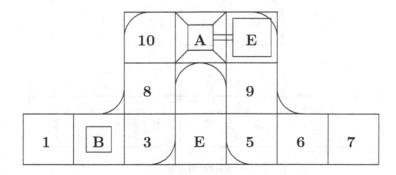

Figure 4.16

Then E decouples from A, moves to 10 and recouples with A, moves to 3 and couples with B, arriving at the position shown in Figure 4.17.

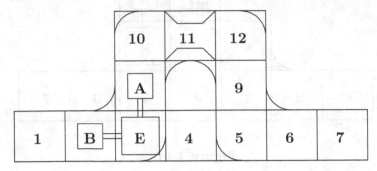

Figure 4.17

Now E moves to 2 in order to back up to 3, arriving at the position shown in Figure 4.18.

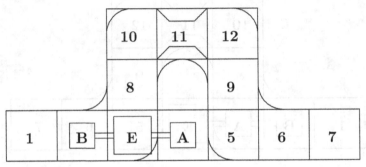

Figure 4.18

Next E decouples from B and moves to 5, arriving at the position shown in Figure 4.19.

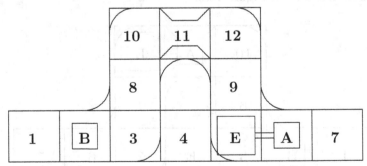

Figure 4.19

Finally, E decouples from A and moves to 4, arriving at the desired position.

Puzzle 5.

Clearly, each of the nyala and the bongo must make at least two moves. If the okapi does not move, then the nyala and the bongo cannot get past each other. Since the okapi eventually gets back to the O-cage, it must also make at least two moves, bringing the total to at least six. Suppose six moves are sufficient. Then each animal makes exactly two moves. Note the obvious symmetry between the N-cage and the B-cage, and the not so obvious symmetry between the O-cage and the courtyard. The first move must be to the courtyard, and there is no point in having the okapi go there. By symmetry we may assume that the first move is made by the bongo. In the second move, the okapi must move to the newly vacated N-cage. In the third move, the bongo cannot move to the O-cage as this means that it will make at least three moves. Hence the nyala moves to the O-cage. Now the bongo moves to the B-cage on the fourth move. Each of the okapi and the nyala has one move remaining, and they are in each other's way. Thus the task cannot be completed in six moves. One more move will suffice since the okapi and the nyala can make a cyclic switch using the courtyard. It follows that the minimum number of moves is seven.

Puzzle 6.

Clearly, each of the nyala and the bongo must make at least two moves. Also, one of the okapi and the yeti must also make at least two moves, bringing the total to at least six. Suppose six moves are sufficient. Then each animal other than either the okapi or the yeti makes exactly two moves. By symmetry, we may assume that the yeti is not moved. Since there is no point in moving either the bongo or the nyala, we may assume that the first move is made by the okapi. In the second move, the bongo or the nyala moves to the newly vacated O-cage and then waits for its correct cage to be vacated. However, whichever of these two that has not moved cannot get to its correct cage. Thus the task cannot be completed in six moves. One more move will suffice. We start with moving the okapi to the left courtyard. Then the bongo goes to the O-cage and onto the right courtyard. The nyala now makes it to the N-cage in two moves. We still have two moves left to get both the okapi and the bongo to their correct cages. It follows that the minimum number of moves is seven.

Puzzle 7.

The moves are shown in Figure 4.20.

    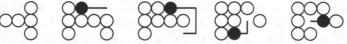

Figure 4.20

Puzzle 8.

The moves are shown in Figure 4.21.

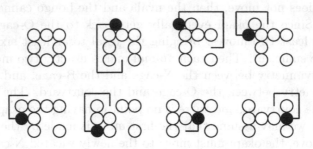

Figure 4.21

Puzzle 9.

Figure 4.22 shows how we can form, in four moves, four stacks of two counters from a row of eight counters. The first two moves are chosen to set up the last two moves.

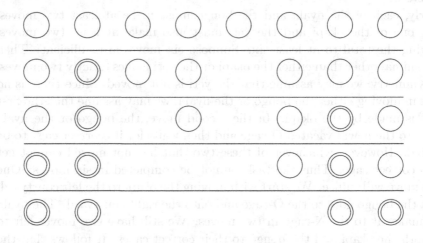

Figure 4.22

It is now a simple matter to form five stacks of two from ten counters in a row. Just jump the third counter from either end of the row on top of the counter at that end. These two counters will not be involved any further. We are now left with eight counters which can be dealt with as before.

Puzzle 10.
Figure 4.23 shows how the task may be accomplished in five jumps.

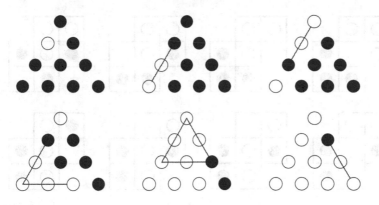

Figure 4.23

Puzzle 11.
Figure 4.24 shows how the task may be accomplished in four jumps.

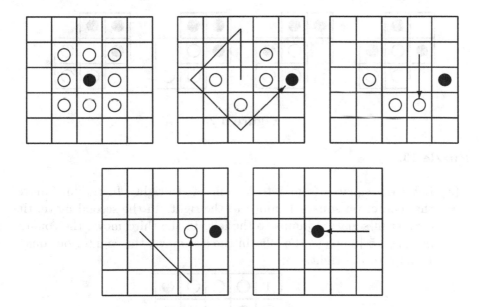

Figure 4.24

Puzzle 12.

Figure 4.25 shows that the task may be accomplished in fifteen moves.

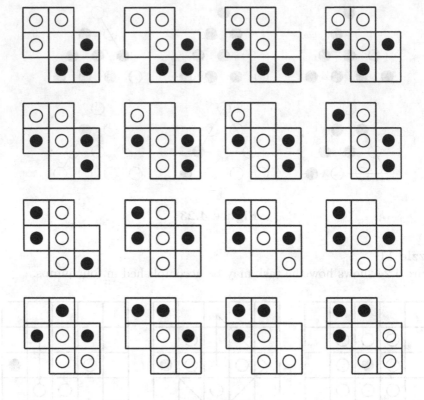

Figure 4.25

Puzzle 13.

(a) Label the squares from 1 to 5 from left to right. In the first move, the counter on square 1 jumps to the right. In the second move, the counter in square 3 jumps to the left. In the third move, the counter in square 5 jumps to the left. In the last move, the counter on square 1 jumps to the right.

1	○	○	○	●
○	●	3	○	●
○	●	●	●	5
1	○	○	○	○

Figure 4.26

(b) The nine moves are shown in Figure 4.27 on the left.

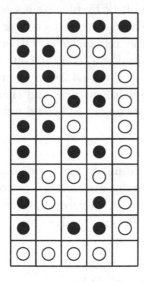 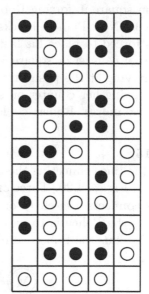

Figure 4.27

(c) The ten moves are shown in Figure 4.27 on the right.

Puzzle 14.

Label the squares as shown in Figure 4.28 on the left. Here is Pamela's winning strategy. She begins by placing a black counter on square 5. Robert has essentially two alternative responses, placing a white counter on a corner square or on an edge square. Suppose Robert places a counter on square 1. Pamela plays her second counter on square 8, forcing Robert to block by placing his second counter in square 2. Now Pamela blocks by placing her last counter on square 3. Robert is forced to place his last counter on square 7. Otherwise, Pamela will move her counter on square 8 there and wins. Now all counters have been placed, arriving at the position shown in Figure 4.28 in the middle. Pamela wins by moving her counters on squares 5 and 8 to squares 6 and 9 respectively in the next two moves. Robert could not stop her, and could not win ahead of her.

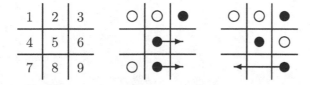

Figure 4.28

47

Suppose Robert places a counter on square 2. Pamela plays her second counter on square 9, forcing Robert to block by placing his second counter in square 1. Now Pamela blocks by placing her last counter on square 3. Robert is forced to place his last counter on square 6. Otherwise, Pamela will move her counter on square 5 there and win. Now all counters have been placed, arriving at the position shown in Figure 4.28 on the right. Pamela wins by moving her counter on square 9 to square 8 and then square 7 in the next two moves. Robert could not stop her, and could not win ahead of her.

Puzzle 15.

Without the line connecting the circles N and R, the board may be redrawn as shown in Figure 4.29. Three of the circles are changed to doubled circles. Each line connects a circle of one type to a circle of another type. At the point when the Police is to move, the Robber is always on a circle of the same kind. Thus the Police cannot possibly win.

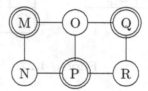

Figure 4.29

In order to win in the given board, Pamela must move the Police once between N and R, and prevent the Robber from doing the same. She has a sure win because this can be accomplished as follows. In the first move, the Robber must move to Q and the Police gives chase at R. In the second move, the Robber must move to O and the Police backs off to N. In the third move, the Robber must move back to Q, and the Police advances to P. In the fourth move, whether the Robber goes to O or R, the Police can land on him.

Chapter Five
Combinatorial Puzzles

Puzzle 1.
Figure 5.1 shows two copies of each of the numbers from 1 to 3 arranged in a row such that
(1) there is exactly 1 other number between the pair of 1s;
(2) there are exactly 2 other numbers between the pair of 2s;
(3) there are exactly 3 other numbers between the pair of 3s.

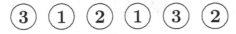

Figure 5.1

Arrange two copies of each of the numbers from 1 to 4 in a row such that
(1) there is exactly 1 other number between the pair of 1s;
(2) there are exactly 2 other numbers between the pair of 2s;
(3) there are exactly 3 other numbers between the pair of 3s;
(4) there are exactly 4 other numbers between the pair of 4s.

Puzzle 2.
Place each of the numbers from 1 to 8 in one of the circles in Figure 5.2, so that the numbers in two circles joined directly by a segment are not consecutive.

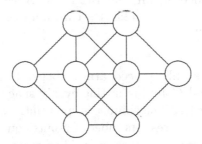

Figure 5.2

Puzzle 3.
Alice, Brian, Colin, Debra and Edwin live on a street with six houses numbered 1 to 6. Each of them lives in a different house, and number 5 is unoccupied. The difference between the house numbers of Alice and Brian is 2 while the sum of the house numbers of Colin and Debra is 7. In which house does Edwin live?

A. Liu et al., *The Puzzles of Nobuyuki Yoshigahara*, Problem Books in Mathematics,
https://doi.org/10.1007/978-3-030-62896-3_5

Puzzle 4.

The following chart shows the answers given by Aaron, Betty, Clara and David in a ten-question true-or-false test. Aaron has eight correct answers, Clara seven and Betty only two. How many correct answers does David have?

Questions	1	2	3	4	5	6	7	8	9	10
Aaron	T	F	F	T	F	F	T	T	F	T
Betty	F	T	F	F	F	T	F	T	F	F
Clara	T	F	T	T	T	F	T	T	T	T
David	F	F	T	F	T	F	F	F	T	F

Puzzle 5.

A girl and a boy played Rock, Scissors and Paper ten times, where Rock beat Scissors, Scissors beat Paper and Paper beat Rock. The boy used Rock three times, Scissors six times and Paper once. The girl used Rock twice, Scissors four times and Paper four times. None of the ten games was a tie. Who won more games?

Puzzle 6.

The numbers from 1 to 9 are written separately on nine cards. Each of Alice, Brian and Colin draws three of the cards, and the sum of the numbers of their cards are the same. A game involves two of them at a time, each putting one card down. They are turned over simultaneously, and the player who has played the card with the higher number wins. It turns out that the game favours Brian over Alice, Colin over Brian but Alice over Colin. If Brian has the card number 1 and Colin has the card number 2, what are the cards drawn by each player?

Puzzle 7.

A spinner has three evenly spaced arms labeled 1, 2 and 3 in clockwise order. Alice, Brian and Colin are sitting evenly spaced around the spinner in counterclockwise order. The spinner is spun until each arm points at one player, and each player scores a number of points equal to the label on that arm. The spinner is spun 6 times, and Alice scores a total of 13 points. Neither Brian nor Colin scores 13 points in total. Who scores the highest number of points?

Puzzle 8.

There are two red marbles and two green marbles in a bag. If you draw two of them out at random, is it more likely for them to be of the same color or of different colors?

Puzzle 9.
Lacy has a necklace with five pearls of different colors, as shown in Figure 5.3, where R stands for red, Y for yellow, B for blue, G for green and W for white. Nick gives her a box for it, and labels the five spaces for the pearls with the five colors. However, no matter how Lacy puts her necklace in the box, exactly one pearl is in the correctly labeled space. How are the spaces labeled?

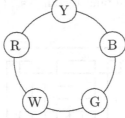

Figure 5.3

Puzzle 10.
A 4 × 15 rectangle is constructed from twelve pieces known as the pentominoes, as shown in Figure 5.4. It is desired to paint each of the pentomino in red, yellow or green, so that no two pentominoes sharing a common segment have the same color. The U-pentomino at the bottom left corner is painted red while the V-pentomino at the top right corner is painted green. How should the remaining pentominoes be painted?

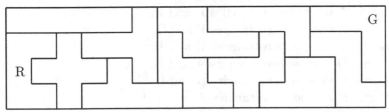

Figure 5.4

Puzzle 11.
In a 4 × 4 table, there is a symbol in each square: a Spade, a Heart, a Diamond or a Club. There is exactly one symbol of each kind in each row, each column and each of the two long diagonals. The symbols in four of the squares are shown in Figure 5.5. Fill in the remaining squares.

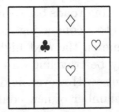

Figure 5.5

51

Puzzle 12.
Place three white queens and five black queens on a 5 × 5 chessboard so that no queen is attacked by another queen of the other color. There is at most one queen on each square, and each queen attacks along her row, her column and her diagonals.

Puzzle 13.
Figure 5.6 shows nine overlapping rectangles. Their labels, A, B, C, D, E, F, G, H and I, have fallen off.

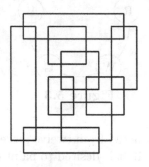

Figure 5.6

It is known that
(1) rectangle A intersects rectangles D and F;
(2) rectangle B intersects rectangles F and G;
(3) rectangle C intersects rectangles G and H;
(4) rectangle D intersects rectangles A and H;
(5) rectangle E intersects rectangles H and I;
(6) rectangle F intersects rectangles A, B and I;
(7) rectangle G intersects rectangles B, C and I;
(8) rectangle H intersects rectangles C, D and E;
(9) rectangle I intersects rectangles E, F and G.
Put the labels correctly back on the rectangles.

Puzzle 14.
In a certain part of a country, there are five parallel highways on which traffic runs from north to south. To squeeze revenue from the motorists, the government constructs connecting roads between the highways, as shown in Figure 5.7. When a car comes to a connecting road, it must turn into it, pay a toll charge, exit and continue on the next highway. The intention is that every motorist will eventually be traveling on the same highway as before entering this part of the country. However, it does not turn out to be the case. What is the minimum number of new connecting roads that must be constructed in order to rectify the situation?

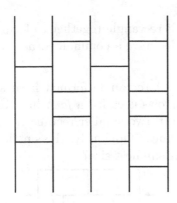

Figure 5.7

Puzzle 15.
Every pair of six cities is to be linked by a highway. No three highways come together except at the cities, but when two of them meet other than at the cities, overpasses are to be constructed, which are expensive. What is the minimum number of overpasses required?

Puzzle 16.
An archaeologist found a Mayan dungeon in the shape of a square divided into four quadrants. Each contained a network of interconnected underground pathways. She constructed maps of each quadrant. After returning from the expedition, she realized that she had forgotten to label the quadrants. So she put them together as shown in Figure 5.8, but she was certain that the pathways did not form just one closed loop. How many closed loops could the pathways form?

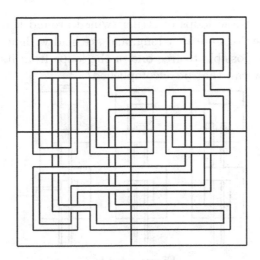

Figure 5.8

Puzzle 17.
Put 32 copies of a 1×3 rectangle together to form a 12×8 rectangle such that no four of the copies have a common corner.

Puzzle 18.
Figure 5.9 shows a 3×4 rectangle formed from six dominoes. The grid line indicated by the arrows is called a *fault* line since the rectangle can be pulled apart along it into two smaller rectangles. A *fault-free* rectangle is one without any fault lines. Obviously, the smallest fault-free rectangle is 1×2. What is the next smallest size?

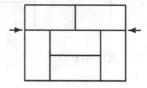

Figure 5.9

Puzzle 19.
Figure 5.10 shows four pairs of intertwining loops. Which of them consists of two unlinked loops?

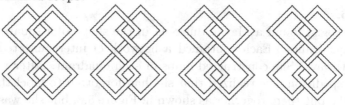

Figure 5.10

Puzzle 20.
Figure 5.11 shows three loops of string twined around three frames. The loose ends of the strings are very long, and are held without being released. Nevertheless, it is possible to free one of the loops from its frame. Which loop is it, and how can that be accomplished?

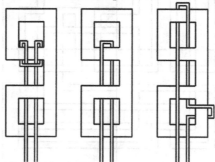

Figure 5.11

Solutions

Puzzle 1.
One of the two copies of 3 must lie between the two copies of 4, and the other outside. The one in between cannot be next to a 4, as otherwise the row will be too long. Hence the one outside is next to a 4. Now there are three spaces between the two copies of 4. If both copies of 2 are in between, then we cannot have exactly 1 other number between the pair of 1s. Hence the 3 between the pair of 4s is also between the pair of 1s. The completed row is shown in Figure 5.12.

Figure 5.12

Puzzle 2.
Each of the two central circles has six neighbors, so that they must be occupied by 1 and 8. The numbers at the ends of the central row must be 2 and 7, with 7 next to 1 and 2 next to 8. Now 3 cannot be next to 2, while 6 cannot be next to 7. We cannot have 3 and 6 next to each other, as that will force 4 and 5 to be next to each other. Since 3 cannot be next to 4 and 6 cannot be next to 5, the placement is unique up to symmetry, as shown in Figure 5.13.

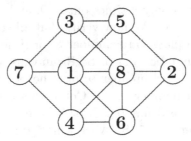

Figure 5.13

Puzzle 3.
Since number 5 is unoccupied, either Alice and Brian live in numbers 1 and 3, or one of them lives in number 4. Hence Colin and Debra cannot live in numbers 3 and 4, and must live in numbers 1 and 6. Then Alice and Brian must live in numbers 2 and 4, leaving number 3 for Edwin.

Puzzle 4.
Call Questions 3, 5 and 9 hard, and the other seven questions easy. Note that Aaron and Clara answer the same way in each of the easy questions but differently in each of the hard questions.

Since Aaron has one correct answer more than Clara, he must have answered correctly two of the three hard questions. Since Betty answers the hard questions the same way as Aaron, she must have answered two of them correctly too. Since she has only two correct answers overall, her answers for each of the easy questions must be wrong! Now David answers the hard questions the same way as Clara. So he gets one of them right. Comparing his answers with Betty's on the easy questions, they differ in Questions 2, 6 and 8. Hence David has answered them correctly, bringing his total to four.

Puzzle 5.

Scissors were used ten times altogether. Since there were no tied games, exactly one player used Scissors in each game. In the six games where the boy used Scissors, the girl won two of them when she used Rock, and lost the other four. In the four games where the girl used Scissors, the boy won three of them when he used Rock, and lost the other one. Hence the boy won seven games and lost only three.

Puzzle 6.

Since $1+2+3+4+5+6+7+8+9=45$, the sum of the numbers on the cards drawn by each player is $45 \div 3 = 15$. Since Brian has card number 1, his other two cards are either numbers 5 and 9 or numbers 6 and 8. In the former case, Colin's cards must be numbers 2, 6 and 7, while Alice's are numbers 3, 4 and 8. In the latter case, Colin's cards must be numbers 2, 4 and 9, while Alice's are 3, 5 and 7. In the former case, Brian beats Alice if he plays card number 9, or if he plays card number 4 while Alice plays card number 3. So Brian wins only 4 times out of 9, and the game favors Alice. Hence the latter is the case. We now verify that all conditions are satisfied. Brian beats Alice if he plays card number 8, or if he plays card number 6 while Alice plays card number 3 or 5. So Brian wins 5 times out of 9. Colin beats Brian if he plays card number 9, or if he plays card number 2 or 4 while Brian plays card number 1. So Colin wins 5 times out of 9. Alice beats Colin if he plays card number 2, or if he plays card number 4 while Alice plays card number 5 or 7. So Alice wins 5 times out of 9.

Puzzle 7.

Note that if Alice scores 1 point, then Brian scores 3 and Colin 2. If Alice scores 2 points, then Brian scores 1 and Colin 3. If Alice scores 3 points, then Brian scores 2 and Colin 1. Alice can score 13 points in three different ways, namely $3+3+3+2+1+1$, $3+3+2+2+2+1$ and $3+2+2+2+2+2$. In the first way, Brian scores $2+3+3+3+1+1=13$ points also. In the second way, Colin scores $1+1+3+3+3+2=13$ points also. Hence it is the third way. Now Brian scores $2+1+1+1+1+1=7$ points and Colin scores $1+3+3+3+3+3=16$ points. It follows that Colin scores the highest number of points.

Puzzle 8.

It does not matter if we draw the two marbles one at a time. The first one is either red or green. Again, it does not matter which because of symmetry. Of the remaining three marbles in the bag, only one matches the color of the marble already drawn. Thus it is twice as likely for the two marbles to be of different colors as for them to be of the same color.

Puzzle 9.

The cyclic order is important, but it cannot remain intact as all five pearls may be in their correctly labeled spaces. So we skip over one pearl at a time and arrive at the labeling shown in Figure 5.14.

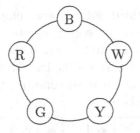

Figure 5.14

If we put the necklace in so that the red pearl is in the space labeled Red, it is the only pearl in its correctly labeled space. If we rotate the necklace so that the red pearl is in the space labeled Blue, the green pearl is the only one in its correctly labeled space. If the red pearl is in the space labeled White, the yellow pearl is the only one in its correctly labeled space. If the red pearl is in the space labeled Yellow, the white pearl is the only one in its correctly labeled space. If the red pearl is in the space labeled Green, the blue pearl is the only one in its correctly labeled space. Now we flip the necklace over. Once again, we can verify that in each of the five positions, exactly one pearl, and a different one each time, is in its correctly labeled space.

Puzzle 10.

The X-pentomino and the I-pentomino are next to each other and both are next to the red U-pentomino. Hence one of them is yellow and the other is green. Denote for now these colors by 1 and 2. Then the T-pentomino must be painted in red, the Y-pentomino in color 2, the W-pentomino in color 1, the N-pentomino in color 2, and F-pentomino in red, the L-pentomino in color 1, the P-pentomino in color 2 and the Z-pentomino in red, as shown in Figure 5.15. Since the L-pentomino is next to the red V-pentomino, color 1 is yellow and color 2 is green.

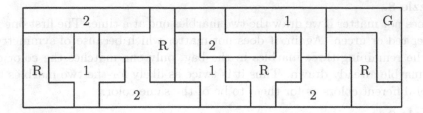

Figure 5.15

Puzzle 11.

The second row and the third column are completely determined by the given data. Suppose the top right corner square contains a Spade. This will then determine completely the first row and the fourth column, leading to the solution shown in Figure 5.16 on the left. If the top right corner square does not contain a Spade, then it must contain a Club. This will lead to the solution in Figure 5.16 on the right.

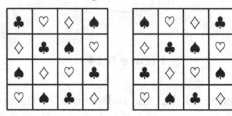

Figure 5.16

Puzzle 12.

We should bunch up queens of the same color as much as possible, and utilize all squares on the chessboard. This suggests placing two of the white queens as shown in the diagram below on the left. The squares marked with crosses are under their attack. This leaves only six squares for the five black queens, and there is still one more white queen to be placed. There is only one square already marked with a cross that attacks only one of those six squares. By placing the third white queen there, we obtain the placement shown in Figure 5.17 on the right.

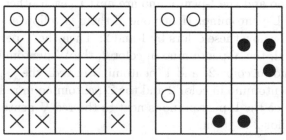

Figure 5.17

Puzzle 13.

Each of the four shaded rectangles in Figure 5.18 intersects three other rectangles, and are therefore F, G, H and I. Now H does not intersect the other three while I intersects both F and G. Hence we can label H and I correctly. The rectangle which intersects both H and I is E. The other rectangle which intersects both F and G is B. Now C intersects both G and H. This allows us to identify C, G and therefore F. Of the remaining two rectangles, the one which intersects F is A and the one which intersects H is D.

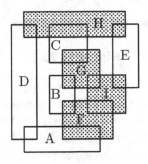

Figure 5.18

Puzzle 14.

Label the connecting roads as shown in Figure 5.19. Roads 1 and 2 cancel each other, as do roads 3 and 4. After that, roads 5 and 6 cancel each other, as do roads 7 and 8. Finally, roads 9 and 10 cancel each other, leaving only road 11. To rectify the situation, we only need to add one connecting road, anywhere between highways 2 and 3. For instance, it may be added where the doubled line is in the diagram below.

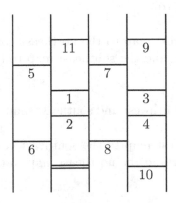

Figure 5.19

Puzzle 15.

Figure 5.20 on the left shows that when there are only four cities, no overpasses are required. The highways divide the plane into four regions, including the outside region. If a fifth city is added, it can be in any one of the regions. It will have direct access to three of the four old cities, so that one overpass is required. When two new cities are added, we have two cases.

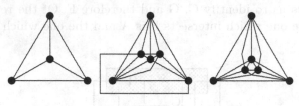

Figure 5.20

Case 1. The new cities are in different regions.

One overpass is needed for each of them to reach the old city not in the region. Moreover, they need another overpass to reach each other. The total is three, as shown in Figure 5.20 in the middle.

Case 2. Both new cities are in the same region.

One overpass is needed for each of them to reach the old city not in the region. Moreover, within that region, when one new city is connected to the three old cities, the region is subdivided into three triangles. When the other new city is added, it will require an overpass to reach one of the three old cities of the region. The total is three, as shown in Figure 5.20 on the right.

In summary, three overpasses are minimum.

Puzzle 16.

If we keep the northwest map in place, the other three maps may be rearranged in five ways, in each of which the pathways form a different number of closed loops, from 2 to 6.

Case 1.

If we move the southwest map to the northeast corner, the northeast map to the southeast corner and the southeast map to the southwest corner, we have 2 closed loops.

Case 2.

If we interchange the southwest and southeast maps, we have 3 closed loops.

Case 3.

If we move the southwest map to the southeast corner, the southeast map to the northeast corner and the northeast map to the southwest corner, we have 4 closed loops.

Case 4.

If we interchange the southwest and northeast maps, we have 5 closed loops.

Case 5.

If we interchange the northeast and southeast maps, we have 6 closed loops.

Puzzle 17.

Let us focus on a 6 × 4 rectangle first. This is easily reassembled with the given condition, as shown at the lower left corner in Figure 5.21 on the left. If we flip this rectangle over and attach it to the right, we have the bottom half of the target rectangle. We now add the shaded part on top to satisfy the given condition, and it is easy to fill in the remaining spaces. A slight modification yields the placement shown in Figure 5.21 on the right.

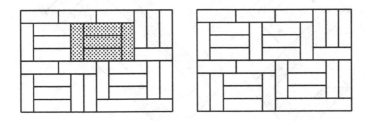

Figure 5.21

Puzzle 18.

A rectangle of width 1 or 2 consisting of at least two dominoes has obvious fault-lines. Consider a rectangle of width 3. We begin tiling it from the left with two dominoes, as shown in Figure 5.22 on the left. In order that the second vertical grid line should not become a fault-line, the next placement of the next three dominoes is forced, as shown in Figure 5.22 on the right. Now we are back to the same situation as before. When we reach the right end of the rectangle, the upper horizontal grid line becomes a fault line. Thus there are no fault-free rectangles of width 3. The same reasoning shows that there are no fault-free rectangles of width 4.

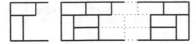

Figure 5.22

It follows that the minimum width of a fault-free rectangle is 5. Since its area is even, the smallest possible rectangle is 5 × 6, using 15 dominoes. There are two such rectangles, as shown in Figure 5.23.

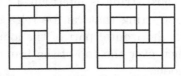

Figure 5.23

Puzzle 19.

Figure 5.24 on the left shows the left loop of a pair. The points A, B, C and D are where it crosses over with the right loop. Reflect the shaded portion about the dotted line, so that the loop changes from figure 8 to figure 0, as shown in Figure 5.24 on the right.

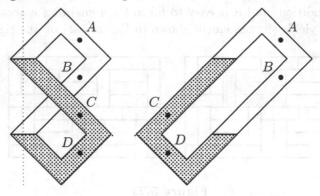

Figure 5.24

The following chart records whether the right loop passes over or under the left loop at points A, B, D and C. Note that because of the reflection, the status at points D and C are reversed.

Pairs	A	B	D	C
First	over	under	over	under
Second	under	over	over	under
Third	under	over	under	over
Fourth	under	over	under	over

In the first, third and fourth pairs, the right loop passes alternately over and under the left loop at A, B, D and C. Hence the two loops cannot be untangled. This is not the case in the second loop. Thus it can be separated into two unlinked loops.

Puzzle 20.

The second string is clearly intertwined with the top part of its frame and cannot be released. The third string is intertwined with both parts of its frame. Hence it is the first string that can be released. This can be accomplished by bringing the horizontal part of the loop down and through the bottom frame, and then over the top of the frame. It is best seen by constructing a physical model.

Chapter Six
Digital Puzzles

Puzzle 1.
Is it possible for the cube of a positive integer to be a four-digit number and the fourth power of the same integer to be a six-digit number?

Puzzle 2.
If the two digits of the father's age are reversed, we get the son's age. The next day, the father will be twice as old as the son. How old is the son today?

Puzzle 3.
Find a number consisting of ten different digits such that the number formed of its first digit is divisible by 1, the number formed of its first two digits is divisible by 2, the number formed of its first three digits is divisible by 3, and so on.

Puzzle 4.
A Yoshigahara multiple of two non-zero single-digit numbers is a positive integer which consists only of these two digits, contains each of them at least once, and is divisible by both numbers. Note that 5 has no Yoshigahara multiple with 2, 4, 6 or 8. Among the least Yoshigahara multiples of all other pairs of single-digit numbers, which is the largest?

Puzzle 5.
Find a number such that if its last digit is removed and put back in as its first digit, the new number is equal to the old number multiplied by

(a) 4;

(b) 6.

Puzzle 6.
The calculator uses 7 bars to represent each of the ten digits, as shown in Figure 6.1. How many two-digit numbers look the same on the display of a calculator when it is held upside down? We do not count 00 since it is not a two-digit number. We count 11 even though the 1s are shifted horizontally.

Figure 6.1

A. Liu et al., *The Puzzles of Nobuyuki Yoshigahara*, Problem Books in Mathematics,
https://doi.org/10.1007/978-3-030-62896-3_6

Puzzle 7.
A faulty calculator does not display any of the vertical bars. One of the digits 2, 3, 5, 6, 8 and 9 has been entered, and only three horizontal bars are displayed. Press an operation sign, then enter a second one-digit number and determine the first number from the result of the operation.

Puzzle 8.
A faulty calculator does not display any of the vertical bars. Figure 6.2 on the left shows a number being entered. Figure 6.2 in the middle shows another number, being entered after the multiplication sign has been pressed. Figure 6.2 on the right shows the product of the two numbers entered. Write down, with restored digits, the equation of this multiplication. Find three answers.

Figure 6.2

Puzzle 9.
Put each of five consecutive digits into a different box in each of the equations in Figure 6.3 to make it correct. Different blocks of consecutive digits may be used. *To eliminate more answers arising from symmetry, we adopt the convention that in an addition or a multiplication, the numbers involved are arranged in descending order.*

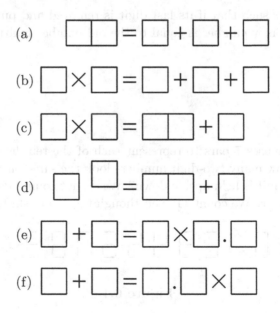

Figure 6.3

Puzzle 10.

Place each of the numbers 1, 2, 3, 4, 5, 6, 7, 8 and 9 once into the nine spaces in the two-part equation shown in Figure 6.4, to make it correct. Two of the numbers have already been placed.

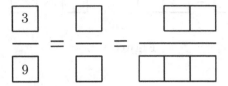

Figure 6.4

Puzzle 11.

(a) Put each of the digits 1, 2, 3, 4 and 5 into a different box in the first equation in Figure 6.5 to make it correct.

(b) Put each of the digits 1, 2, 3, 4, 5 and 6 into a different box in the second equation in Figure 6.5 to make it correct.

(c) Put each of the digits 1, 2, 3, 4, 5, 6, 7 and 8 into a different box in the third equation in Figure 6.5 to make it correct.

(d) Put each of the digits 1, 2, 3, 4, 5, 6, 7, 8 and 9 into a different box in the fourth equation in Figure 6.5 to make it correct.

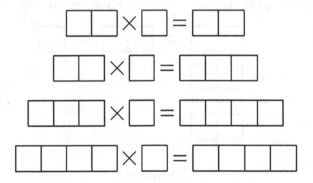

Figure 6.5

Puzzle 12.
Put each of the digits 1, 2, 3, 4, 5, 6, 7, 8 and 9 into a different box in the two equations in Figure 6.6 to make both of them correct.

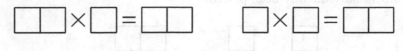

Figure 6.6

Puzzle 13.
Place each of the numbers 1, 2, 3, 4, 5, 6, 7, 8 and 9 once into the nine spaces in the two-part equation shown in Figure 6.7 to make it correct.

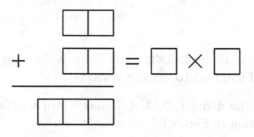

Figure 6.7

Puzzle 14.
Place each of the numbers 1, 2, 3, 4, 5, 6, 7, 8 and 9 once into the nine spaces in the two-part equation shown in Figure 6.8 to make it correct.

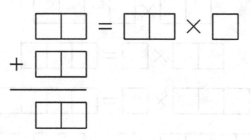

Figure 6.8

Puzzle 15.
Put each of the digits 1, 2, 3, 4, 5, 6, 7, 8 and 9 into a different box in the equation in Figure 6.9 to make it correct.

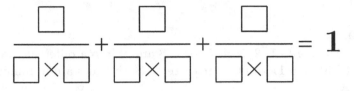

Figure 6.9

Puzzle 16.
Put each of the digits 1, 2, 3, 4, 5, 6, 7, 8 and 9 into a different box in the equation in Figure 6.10 to make it correct.

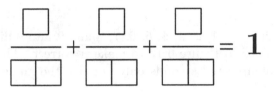

Figure 6.10

Puzzle 17.
A 24-hour clock shows 1 or 2 digits for the hour, 2 digits for the minute, and 2 digits for the second, separated by colons. Mary's also shows 1 digit for the month and 2 digits for the day in front of the time, and the hour is always shown as 2 digits. A time such as 8:19:23:46:57 is *pandigital* since it displays each of the nine non-zero digits exactly once, and not 0. In a year, starting from 1:01:00:00:00 to 12:31:23:59:59, there are 768 occasions when the time is pandigital.

(a) Find the earliest pandigital time in a year.

(b) Find the latest pandigital time in a year.

Puzzle 18.
A time such as 3:59:53 is *palindromic*, that is, it reads the same forward and backward. In a day, starting from 0:00:00 to 23:59:59, there are 660 occasions when the time is palindromic.

(a) Find the two palindromic times that are closest to each other.

(b) Find the two palindromic times that are farthest apart with no other palindrome time in-between.

(c) Find the two palindromic times that are farthest apart from each other. These are *not* the earliest palindromic time 0:00:00 and the latest palindromic time 23:55:32 since they are only 4 minutes and 28 seconds apart.

Puzzle 19.
Put each of the digits 1, 2, 3, 4, 5, 6, 7, 8 and 9 into a different box in the equation in Figure 6.11 to make it correct. Both times are in minutes and seconds only.

Figure 6.11

Puzzle 20.
Put each of the digits 0, 1, 2, 3, 4, 5, 6, 7, 8 and 9 into a different box in each of the equations in Figure 6.12 to make it correct. The first time in each part is in minutes and seconds only. In (b), the minute is shown as only 1 digit.

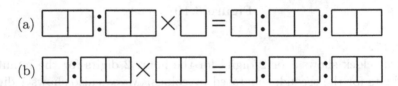

Figure 6.12

Solutions

Puzzle 1.
When the fourth power is divided by the third power, the quotient is greater than $\frac{100000}{10000} = 10$. Testing numbers starting from 11, we find that $18^3 = 5832$ while $18^4 = 104976$. We may also use 19, 20 or 21 instead of 18.

Puzzle 2.
Let x and y be the digits of the father's age which is $10x + y$. The son's age is $10y + x$, and $y < x$. There are four cases to consider.
Case 1. Tomorrow will be the birthday of neither the father nor the son.
Then $10x + y = 2(10y + x)$, which simplifies to $8x - 19y = 0$. The smallest positive integral value for x is 19, but this contradicts the condition that x is a digit. Hence this case is impossible.
Case 2. Tomorrow will be the birthday of both the father and the son.
Then $10x + y + 1 = 2(10y + x + 1)$, which simplifies to $8x - 19y = 1$. By inspection, the smallest positive integral value for x is 12. Thus this case is also impossible.
Case 3. Tomorrow will be the birthday of the father but not the son.
Then $10x + y + 1 = 2(10y + x)$, which simplifies to $8x - 19y = -1$. By inspection, we have $x = 7$ and $y = 3$. So the father is 73 and the son is 37.
Case 4. Tomorrow will be the birthday of the son but not the father.
Then $10x + y = 2(10y + x + 1)$, which simplifies to $8x - 19y = 2$. By inspection, we have $x = 5$ and $y = 2$. So the father is 52 and the son is 25.
In summary, the son's age is either 25 or 37.

Puzzle 3.
Let us call the number formed of the first digit the first number, the number formed of the first two digits the second number, and so on. The prime numbers under ten are 2, 3, 5 and 7. Let us first use the prime 5. Since the tenth number is divisible by 10, the last digit must be 0. Since the fifth number is divisible by 5, the fifth digit must be 5. Next, we use the prime 3. The ninth number is automatically divisible by 9. The sum of the first three digits is a multiple of 3. The sum of the first six digits is also a multiple of 3. Hence the sum of the middle three digits is a multiple of 3. We now use the prime 2. The digits 2, 4, 6 and 8 must be in the second, fourth, sixth and eighth places, each preceded by an odd digit. Since the fourth and the eighth numbers are divisible by 4, the digits in the fourth and the eighth places are 2 and 6. We consider two cases.
Case 1. The fourth digit is 2 and the eighth digit is 6.
Then the sixth digit must be 8 since 2+5+4 is not divisible by 3. Now the second digit is 4. Then the seventh and the ninth digits must be 3 and 9 as neither can be the first or the third digit to go with 4. Since 836 is not divisible by 8, the seventh digit is 9. However, neither 1472589 nor 7412589 is divisible by 7.

Case 2. The fourth digit is 6 and the eighth digit is 2.
Then the sixth digit must be 4 since 6+5+8 is not divisible by 3. Now the second digit is 8. The seventh digit is not 1 or 9 as neither 612 nor 692 is divisible by 8. Of the numbers 1836547, 3186547, 1896547, 9186547, 1896543, 9186543, 7896543 and 9876543, only the second one is divisible by 7.
It follows that the desired ten-digit number is 3816547290.

Puzzle 4.
We first compute the least Yoshigahara multiple of 7 and 9. In order to be divisible by 9, its digit-sum must be divisible by 9. Since we must have at least one copy of 7, we must have at least nine copies of it. In order for the number to be divisible by 7, the digits which are 7 do not matter. The residual number consists of 9s and 0s. Since 9 and 7 are relatively prime, we seek the smallest multiple of 7 which consists of 1s and 0s, and this number is 1001. Around 9009, we add nine copies of 7, yielding the eleven-digit number 77777779779. The following chart shows the least Yoshigahara multiple between any pair of non-zero single-digit numbers. None of them is larger than 77777779779.

1,2	12	3,7	37737
1,3	1113	3,8	3888
1,4	144	3,9	3339
1,5	15	4,5	none
1,6	1116	4,6	4464
1,7	1771	4,7	44744
1,8	1888	4,8	48
1,9	1111111119	4,9	4444444944
2,3	2232	5,6	none
2,4	24	5,7	5775
2,5	none	5,8	none
2,6	2226	5,9	5555555595
2,7	2772	6,7	66766
2,8	288	6,8	6888
2,9	2222222292	6,9	6696
3,4	3444	7,8	7888888
3,5	3555	7,9	77777779779
3,6	36	8,9	8888889888

Puzzle 5.

(a) A young student may approach the problem this way. The digit moved is at least 4. Suppose it is 4. We perform the following divisions. We simply add the preceding quotient to the dividend, until we arrive at an exact division with 4 as the last digit of the quotient.

$$\frac{1}{4)4} \quad \frac{10}{4)41} \quad \frac{102}{4)410} \quad \frac{1025}{4)4102} \quad \frac{10256}{4)41025} \quad \frac{102564}{4)410256}$$

We have $102465 \times 4 = 410256$. If the first digit is 5, 6, 7, 8 or 9, we get different answers: $4 \times 128205 = 512820$, $4 \times 158346 = 615834$, $4 \times 179487 = 717948$, $4 \times 205128 = 820512$ and $4 \times 230769 = 923076$. Here is a more sophisticated approach. Let $d \geq 4$ be the last digit of the desired positive integer. Create a pure repeating decimal x with the desired positive integer as the repeating block. Then $40x$ is exactly the same as x with an extra d in front of the decimal point. It follows that $x = \frac{d}{39}$. In fact, $\frac{4}{39} = 0.102564102564...$, and so on.

(b) Let the digit moved be $d \geq 6$. Then the desired positive integer is the repeating block of $\frac{d}{59}$. For $d = 6$, it is

101694915254237288135593220338983050847457627118 6440677966.

This is one sixth of

6101694915254237288135593220338983050847457627118644067796.

Different answers may be obtained using 7, 8 or 9 as the moving digit.

Puzzle 6.
To 11, 22, 55 and 88, we must add 69 and 96. So there are six such numbers altogether.

Puzzle 7.
We may press the division sign and then enter the number 2. We may have $2 \div 2 = 1$, $3 \div 2 = 1.5$, $5 \div 2 = 2.5$, $6 \div 2 = 3$, $8 \div 2 = 4$ and $9 \div 2 = 4.5$. These quotients are shown in Figure 6.13, and they are distinguishable from one another.

Figure 6.13

Puzzle 8.

The units digit of the product is 7. Hence the units digits of the multiplicand and the multiplier are 3 and 9. Thus the equation appears to be either $49 \times 3 = 147$ or $43 \times 9 = 387$. The former is rejected since each the hundreds digit and the tens digit of the product must be one of 2, 3, 5, 6, 8 and 9. Additional answers can only exist because there may be invisible 1s in front of any of the three numbers. Testing reveals that the equation may be $143 \times 9 = 1287$ or $49 \times 13 = 637$.

Puzzle 9.

These are all quite easy to find. In (c), another answer is $9 \times 8 = 65 + 7$. In (d), another answer is $3^4 = 75 + 6$.

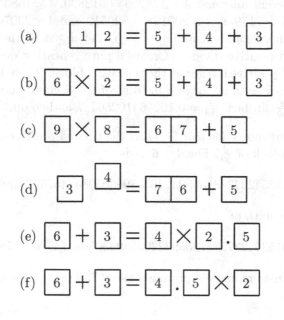

Figure 6.14

Puzzle 10.

The fraction in the middle of the equation must be $\frac{2}{6}$. It cannot be $\frac{1}{3}$ since the 3 has already been used. In the fraction on the right, the hundreds digit of the denominator is clearly 1 since 2 has been used. Now 3 times 4, 5, 7 and 8 end in 2, 5, 1 and 4 respectively. Only the last combination is feasible since 2 and 1 have been used and 5 cannot be used twice. Routine checking reveals that the tens digit of the numerator must be 5, and that of the denominator must be 7. The completed equation is shown in Figure 6.15.

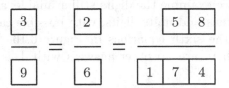

Figure 6.15

Puzzle 11.

(a) Neither 1 nor 5 can appear as a units digit. Hence the multiplication can only be the first equation in Figure 6.16.

(b) Neither 1 nor 5 can appear as a units digit. In fact, 1 must be the hundreds digit in the product. The multiplication can only be the second equation in Figure 6.16.

(c) In problem 10, we see that $58 \times 3 = 174$. The digits still available are 2 and 6, and $2 \times 3 = 6$. This leads to the third equation in Figure 6.16. Computer search reveals that there is the only other answer is the fourth equation in Figure 6.16.

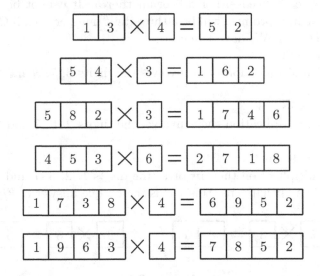

Figure 6.16

(d) In (c), we use the divide-and-conquer method with the multiplier 3. Let us try it again, but with the multiplier 4. Focus on the thousands and hundreds digits. We may have $13 \times 4 = 52$, $17 \times 4 = 68$, $18 \times 4 = 72$ or $19 \times 4 = 76$. We may also have $12 \times 4 = 48$, provided that there is a carry-over of 2, or $23 \times 4 = 92$, provided that there is a carry-over of 3.

In each case, we examine the digits still available, and see if they can complete the tens and units digits, with possible carry-over. We find the fifth and the sixth equations in Figure 6.16. Computer search reveals that there are no other answers with 4 as the multiplier or otherwise.

Puzzle 12.
Label the nine boxes as shown in Figure 6.17.

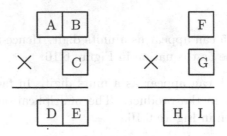

Figure 6.17

Clearly, 1 cannot be B, C, F or G. Now 5 cannot be B, C, E, F, G and I, as otherwise we need either a 0 or another 5. It cannot be A either, as otherwise the first product will be a three-digit number even if C is 2. Hence 5 is either D or H. We have five cases.

Case 1. $52 = 13 \times 4$.
The second multiplication then involves the digits 6, 7, 8 and 9. This will not work.

Case 2. $54 = 18 \times 3$.
The second multiplication then involves the digits 2, 6, 7 and 9. This will not work.

Case 3. $54 = 9 \times 6$.
The first multiplication then involves the digits 1, 2, 3, 7 and 8. We have $27 \times 3 = 81$. As it happens, this yields the unique answer in Figure 6.18.

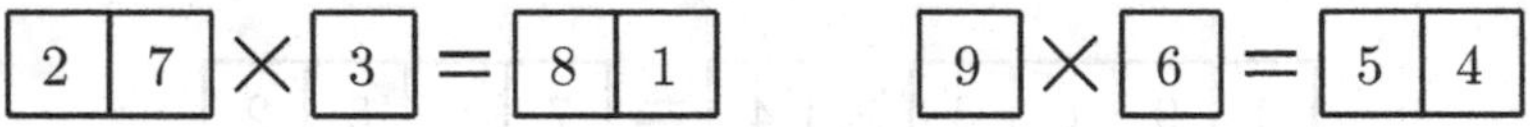

$$\boxed{2}\,\boxed{7} \times \boxed{3} = \boxed{8}\,\boxed{1} \qquad \boxed{9} \times \boxed{6} = \boxed{5}\,\boxed{4}$$

Figure 6.18

Case 4. $56 = 8 \times 7$.
The first multiplication then involves the digits 1, 2, 3, 4 and 9. This will not work.

Case 5. $57 = 19 \times 3$.
The second multiplication then involves the digits 2, 4, 6 and 8. This will not work.

Puzzle 13.

Clearly, the hundreds digit in the sum must be 1. We focus first on the multiplication which involves four different digits. There are ten possibilities. The cases $24 = 8 \times 3$, $27 = 9 \times 3$, $28 = 7 \times 4$, $32 = 8 \times 4$ and $36 = 9 \times 4$ are impossible as the product is too small for the addition to work. We may have $54 = 9 \times 6$ or $63 = 9 \times 7$ as shown in Figure 6.19. The remaining cases, namely, $42 = 7 \times 6$, $56 = 8 \times 7$ and $72 = 9 \times 8$, can be eliminated by routine checking.

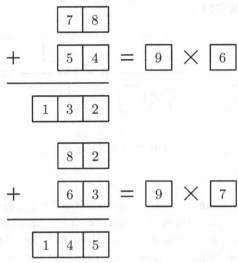

Figure 6.19

Puzzle 14.

We focus first on the multiplication which involves five different digits. There are sixteen possibilities. The cases $98 = 14 \times 7$, $96 = 12 \times 8$, $87 = 29 \times 3$, $84 = 12 \times 7$, $81 = 27 \times 3$, $78 = 13 \times 6$ and $76 = 19 \times 4$ are impossible as the product is too large for the addition to work. We may have $68 = 17 \times 4$ as shown in Figure 6.20. The remaining cases, namely $78 = 26 \times 3$, $72 = 18 \times 4$, $57 = 19 \times 3$, $54 = 18 \times 3$, $52 = 13 \times 4$, $48 = 16 \times 3$, $36 = 18 \times 2$ and $34 = 17 \times 2$, can be eliminated by routine checking.

$$\boxed{6}\,\boxed{8} = \boxed{1}\,\boxed{7} \times \boxed{4}$$
$$+\ \boxed{2}\,\boxed{5}$$
$$\overline{\ \boxed{9}\,\boxed{3}}$$

Figure 6.20

Puzzle 15.

Except for 5 and 7, each of the other seven digits is a divisor of 72. Hence we aim for 72 as the least common denominator of the three fractions. A reasonable approach is to make the first fraction as large as possible, with 7 as its numerator. Hence the denominator is either 8×1 or 4×2. The other two fractions then add up to $1 - \frac{7}{8} = \frac{9}{72}$. The numerator of the second fraction is 5, and its denominator must be 9×8. Hence the denominator of the first fraction is 4×2. The third fraction is $\frac{4}{72} = \frac{1}{6 \times 3}$, leading to the unique answer in Figure 6.21.

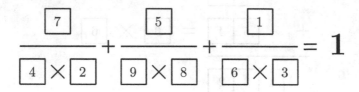

Figure 6.21

Puzzle 16.

One of the three fractions has to be fairly close to 1. The closest we can make it is to use 9 as the numerator and 12 as the denominator. Then the sum of the other two fractions is $\frac{1}{4}$ and they use the digits 3, 4, 5, 6, 7 and 8. The two denominators must have common factors in order for the denominator of the sum to reduce to 4. The obvious choices are 34 and 68, and indeed this leads to the answer in Figure 6.22. Computer search reveals that it is unique.

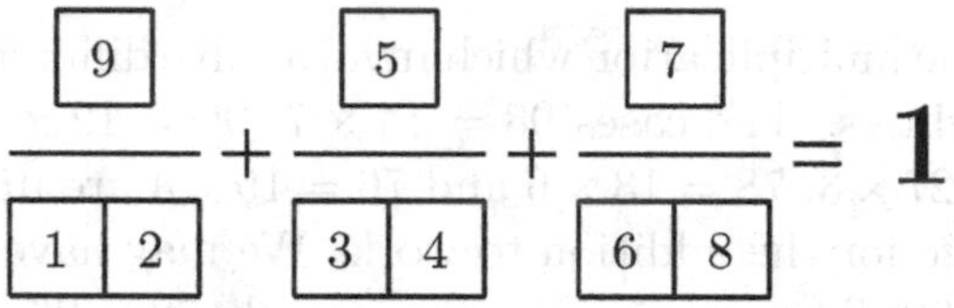

$$\frac{9}{12} + \frac{5}{34} + \frac{7}{68} = 1$$

Figure 6.22

Puzzle 17.

(a) The earliest pandigital time in a year is 3:26:17:48:59. None of 6, 7, 8 and 9 can appear as any of the tens digits. The tens digit of the hour must be 1 because the hour cannot be higher than 24. The tens digit of the day must be 2 as it cannot be higher than 31. Thus the smallest digit for the month is 3. The remaining tens digits are 4 and 5 in that order, while the units digits are 6, 7, 8 and 9 in that order.

(b) The latest pandigital time in a year is 9:28:17:56:43. Reasoning as in (a), the tens digit of the hour is 1 and the tens digit of the day is 2. The other tens digits are 5 and 4 in that order. The units digits are then 9, 8, 7, 6 and 3 in that order.

Puzzle 18.

(a) These are 9:59:59 and 10:00:01, only 2 seconds apart.

(b) These are 15:55:51 and 20:00:02, 4 hours 4 minutes and 11 seconds apart. We use the fact that the hour cannot be 16, 17, 18 or 19 since 61, 71, 81 and 91 are not acceptable numbers for the second.

(c) These are 1:33:31 and 13:33:31, 12 hours apart, which is the maximum possible difference.

Puzzle 19.
Label the nine boxes as shown in Figure 6.23.

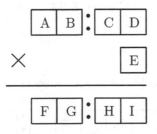

Figure 6.23

Note that none of 6, 7, 8 and 9 can be A, C, F and H. It follows that none of them can be E, either. Now A can only be 1, and since B is at least 6, E can only be 2 or 3. We consider these two cases.
Case 1. E is 2.
Then the carry-over from the second part to the minute part is at most 1. Note that $6 \times 2 = 12$, $7 \times 2 = 14$, $8 \times 2 = 16$ and $9 \times 2 = 18$. Hence D is either 8 or 9.
Subcase 1(a). D is 8.
Then I is 6, so that B and G are 7 and 9. If B is 7, G cannot grow from 4 to become 9. If B is 9, G cannot grow from 8 to become 7. Hence there are no answers here.
Subcase 1(b). D is 9.
Then I is 8, so that B and G are 6 and 7. If B is 6, G cannot grow from 2 to become 7. If B is 7, G still cannot grow from 4 to 6 since the carry-over is at most 1. Hence there are no answers here.

Case 2. E is 3.

Then we may have a carry-over of 2 from the second part to the minute part. Note that $6 \times 3 = 18$, $7 \times 3 = 21$, $8 \times 3 = 24$ and $9 \times 3 = 27$. Hence D is either 6 or 9.

Subcase 2(a). D is 6.

Then I is 8 so that B and G are 7 and 9. If B is 7, G cannot grow from 1 to become 9. If B is 9, G is indeed 7, with no carry-over. This means that F is 5, so that C and H are 2 and 4. However, this does not work. Hence there are no answers here.

Subcase 2(b). D is 9.

Then I is 7 so that B and G are 6 and 8. If B is 6, G is indeed 8, with no carry-over. This means that F is 4, so that C and H are 2 and 5. However, this does not work. If B is 8, G can grow from 4 to 6 with a carry-over of 2. This leads to the unique answer in Figure 6.24.

Figure 6.24

Puzzle 20.

(a) Computer search reveals the unique answer in Figure 6.25.

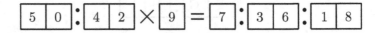

Figure 6.25

(b) Computer research reveals the unique answer in Figure 6.26.

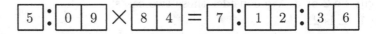

Figure 6.26

Remark:

We do not see how this puzzle can be solved heuristically. It serves as a reminder that once in a while, the machine may have the advantage over the human.

Chapter Seven
Number Puzzles

Puzzle 1.
Put each of the digits 1, 2, 3, 4, 5, 6, 7, 8 and 9 into a different small circle in the Olympic symbol shown in Figure 7.1, so that the digits inside each large circle has the same sum. What are the minimum and maximum values of this constant sum?

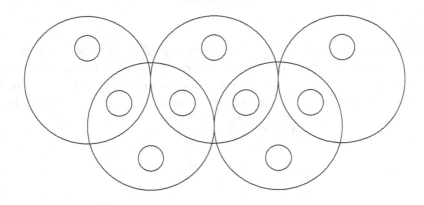

Figure 7.1

Puzzle 2.
Place each of the numbers from 1 to 7 in one of the circles in Figure 7.2, such that the sum of the numbers in the three circles in each of the five lines will be the same. Which of the numbers must go into the circle marked "?"?

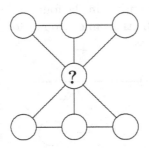

Figure 7.2

Puzzle 3.
Place each of seven of the numbers from 1 to 9 in a different circle in Figure 7.3, such that the product of the numbers in the three circles in each of the three lines will be the same.

79

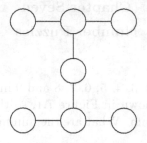

Figure 7.3

Puzzle 4.

Place each of the numbers from 1 to 10 in one of the circles in Figure 7.4, such that the sum of the numbers in the circles along the four sides of the rectangle will be the same. Two of the numbers have been placed for you. In both the top and bottom rows, the second number is larger than the third.

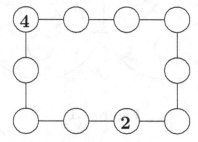

Figure 7.4

Puzzle 5.

Place each of the numbers from 1 to 8 in one of the circles in Figure 7.5, such that the sum of the numbers in the four circles at the corners of each of the six quadrilaterals will be the same. Do not forget the outer quadrilateral which contains the other five.

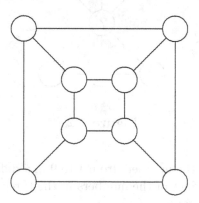

Figure 7.5

Puzzle 6.

Place each of the numbers from 1 to 15 in one of the circles in Figure 7.6, so that the sum of the numbers in two circles joined by a line segment will be the square of an integer.

Figure 7.6

Puzzle 7.

In Figure 7.7, each number not in the top row of an upside down numerical triangle is the difference between the two in the row above that are to its immediate right and immediate left.

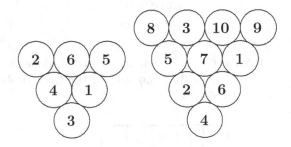

Figure 7.7

(a) Find three more examples for triangles of each size, using the numbers 1 to 6 and 1 to 10, respectively.

(b) Use the numbers 1 to 15 to construct such a triangle of the next size.

Puzzle 8.

Divide the table Figure 7.8 into two connected parts, each consisting of a number of squares, such that the total of the numbers in the squares of each part is the same.

1	2	3	4
5	6	7	8
9	10	11	12

Figure 7.8

Puzzle 9.

Divide the table in Figure 7.9 into two connected parts, each consisting of a number of squares, such that the total of the numbers in the squares of each part is the same.

1	2	3	4
5	6	7	8
9	10	11	12
13	14	15	16

Figure 7.9

Puzzle 10.

Choose five squares from the 5×5 table in Figure 7.10, with one in each row and one in each column, so that the sum of the five numbers in these squares is as large as possible.

16	7	13	9	5
20	10	16	13	9
15	6	10	8	4
21	12	17	14	10
11	2	7	3	0

Figure 7.10

Puzzle 11.

Just before a magic performance, you were called away. You explained to your assistant how she could do a trick. While she was out of the room, the audience covered up one of the numbers of a 10×10 table before bringing her back. She only needed to take one look at the table and told them what the covered-up number was. She had not even seen it before, but the table had a special property that allowed her to perform without failure. What special property does the table in Figure 7.11 have?

5	5	1	6	2	7	3	8	4	9
1	9	5	8	4	7	3	6	2	5
6	4	2	5	3	6	4	7	5	8
2	8	6	7	5	6	4	5	3	4
7	3	3	4	4	5	5	6	6	7
3	7	7	6	6	5	5	4	4	3
8	2	4	3	5	4	6	5	7	6
4	6	8	5	7	4	6	3	5	2
9	1	5	2	6	3	7	4	8	5
5	5	9	4	8	3	7	2	6	1

Figure 7.11

Puzzle 12.

There are nine one-digit numbers, not necessarily distinct, whose sum is 45 and whose product is 362880. Do they have to be 1, 2, 3, 4, 5, 6, 7, 8 and 9?

Puzzle 13.

In a shooting gallery, nine prizes, values \$1 to \$9, were hung up on three strings, as shown in Figure 7.12. There was a target above each prize. If a target was hit, everything below it went to the shooter. Alice shot first, followed by Brian and then Colin. Each shot twice in a row. Alice got \$18 in prizes, Brian \$13 and Colin \$14. Which targets were hit, and by which shooters?

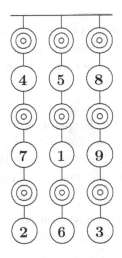

Figure 7.12

Puzzle 14.

The target in the next shooting gallery consisted of four concentric rings around a central circle. A hit in the central circle was worth 37 points. A hit in the rings outward from the central circle was worth 23, 19, 17 and 11 points respectively. Was it possible to score exactly 100 points with four hits?

Puzzle 15.

A sports team has ten players numbered from 1 to 10, and five locker rooms numbered from 9 to 13. Each locker room is being used by two players, and the sum of the numbers of the two players is equal to the number of the locker room. If player number 7 uses locker room number 10, which locker room does player number 4 use?

Puzzle 16.

During renovation, three of the locker rooms are renumbered 5, 7 and 10, while the other two are unnumbered. Each locker room is being used by two players, and the sum of the numbers of the two players is equal to the number of the locker room if it is numbered. Which are the two players in each of the three numbered locker rooms?

Puzzle 17.

A cheetah and a fox are having a race on a straight 100-meter track. The cheetah makes 2 hops per minute and covers 3 meters per hop. The fox makes 3 hops per minute, but covers 2 meters per hop. Neither can turn around before they reach the end of the track. Which animal wins the race by returning first to the starting point?

Puzzle 18.

There are three clocks in the house. One gives the time 4:56 pm, the second 5:01 pm and the third 5:10 pm. They are off by 4, 5 and 10 minutes, not necessarily respectively. What is the correct time?

Problem 19.

The army was affected by a severe budget cut, and only half of the soldiers in each unit had uniforms. When the Emperor came, two units were presented for inspection. All soldiers in the unit formed a rectangular array. The soldiers on the perimeter all had uniforms, but those inside did not. The two units were of different sizes. How many soldiers were in each unit?

Puzzle 20.

There are 50 coins, each either a penny, a nickel or a dime. If the total value is exactly one dollar, how many coins of each kind are there?

Puzzle 21.

A shop accepts bills of 1, 5, 10 and 50 Euros. A customer whose purchase is worth 100 Euros pays for the exact amount with 21 bills. In how many different ways can this be done?

Puzzle 22.
A shopkeeper sets the sales prices of the thirteen volumes of Euclid's *Elements* arbitrarily at $42, $27, $18, $45, $60, $9, $6, $12, $29, $72, $33, $70 and $90 respectively. Which volumes must have been sold if the shopkeeper takes in $100 in total?

Puzzle 23.
A lone tree stands at one end of a highway, along which more trees are to be planted. If the trees are planted 20 meters apart from each other, the total cost is $600. If they are planted 10 meters apart, the total cost is $1,200. What is the total cost if they are planted 15 meters apart?

Puzzle 24.
Three workers are paid to cut a large lawn, but the first worker is unable to come. The second worker clears $\frac{5}{9}$ of the lawn while the third worker clears the remaining $\frac{4}{9}$ of the lawn. Each worker is paid $90. What is the fair way for the second and third workers to share the $90 that is going to the first worker?

Puzzle 25.
A rich man dies and leaves behind five gold bars to his mother, wife and son. The gold bars are of respective weights 4, 9, 11, 12 and 19 kilograms. The man stipulates that the son is to receive twice as much gold as the man's mother. Which gold bar goes to which person?

Puzzle 1.

Note that 1+2+3+4+5+6+7+8+9=45 is a multiple of 5. Four of these nine numbers are used twice. Since there are five large circles, the sum of these four numbers must also be a multiple of 5. The smallest value is 1+2+3+4=10, yielding the minimum constant sum of $(45 + 10) \div 5 = 11$. This can be realized as shown in Figure 7.13.

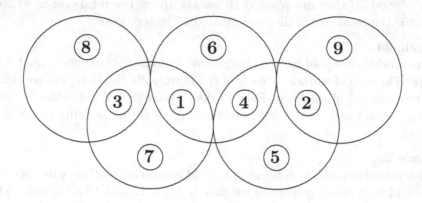

Figure 7.13

The largest sum of four of these nine numbers is 6+7+8+9=30, leading to a potential maximum constant sum of $(45 + 30) \div 5 = 15$. However, with 6, 7, 8 and 9 all in the middle row, there is no way for the sum of the numbers inside each of the large circles at the two ends to reach 15. The maximum value is 14, as realized in Figure 7.14.

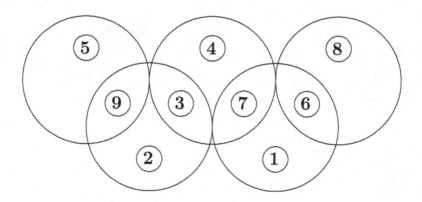

Figure 7.14

Puzzle 2.

When we compute the total of the five sums, each number is counted twice except the one which goes into the circle marked "?". Now $2 \times (1+2+\cdots+7) = 56$. Since the total must be a multiple of 5, this number must be 4, and each sum is $(56+4) \div 5 = 12$. Figure 7.15 shows one of many possible placements of the numbers.

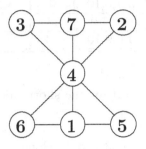

Figure 7.15

Puzzle 3.

Clearly, the numbers to be omitted are 5 and 7. All the others are divisors of 72, which is the only possible choice for the magic constant. Groups of three of the numbers with 72 as product are $1 \times 8 \times 9$, $2 \times 4 \times 9$ and $3 \times 4 \times 6$. They can be assembled as shown in Figure 7.16.

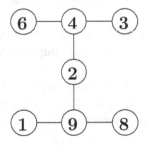

Figure 7.16

Puzzle 4.

The four numbers at the corners are counted twice while every other number is counted only once. The combined total of the four lines must be a multiple of 4. Note that $1+2+\cdots+10 = 55$. Since we must add at least $1+3+4+5=13$ to it, the magic constant is at least $(55 + 13) \div 4 = 17$. On the other hand, we can add at most $4+8+9+10=31$ to it, so that the magic constant is at most $(55 + 31) \div 4 = 21.5$. Since it is an integer, it is at most 21. The following chart lists the thirteen possible cases.

Case #	Magic Constant	Corner Numbers	Case #	Magic Constant	Corner Numbers
1	17	1,3,4,5			
2	18	1,3,4,9	3	18	1,4,5,7
4	19	1,4,6,10	5	19	1,4,7,9
6	19	3,4,5,9	7	19	3,4,6,8
8	20	3,4,8,10	9	20	4,5,6,10
10	20	4,5,7,9	11	20	4,6,7,8
12	21	4,6,9,10	13	21	4,7,8,10

Case 1.
This is impossible as the total for the short side containing 1 is at most
1+5+10=16.
Case 2.
This is impossible as two of 1, 3 and 4 belong to the same short side, and
its total is at most 3+4+10=17.
Case 3.
The short side containing 4 must consist of 4+9+5 and the other short side
of 1+10+7. The long side containing 2 must consist of 5+4+2+7, but 4 is
already used. Thus this case is impossible.
Case 4.
Note that 1 must be on the same short side as 10. The long side containing
2 must also contain 6, along with two numbers with sum 11. Since exactly
one of them has to be 1 or 10, this case is impossible.
Case 5.
The short side containing 1 must also contain 9, but then we need another
copy of 9 in between. Hence this case is impossible.
Case 6.
Two of 3, 4 and 5 belong to the same short side, which must consist of
4+10+5. The long side containing 2 must also contain 5, along with two
numbers with sum 12. Since exactly one of them has to be 3 or 9, this case
is impossible.
Case 7.
Two of 3, 4 and 6 belong to the same short side, which must consist of
4+9+6. The long side containing 2 must also contain 6, along with two
numbers with sum 11. Since exactly one of them has to be 3 or 8, this case
is impossible.
Case 8.
This case yields the solution in Figure 7.17.
Case 9.
The short side containing 4 must contain two other numbers with sum 16.
Hence neither is 5, so that exactly one of them is 6 or 10. Thus this case is
impossible.

Case 10.
The short side containing 4 must contain two other numbers with sum 16. Hence neither is 5, so that exactly one of them is 7 or 9. Thus this case is impossible.

Case 11.
The short side containing 4 consists of either 4+9+7 or 4+10+6. In the former instance, the long side containing 4 must contain 1 and either 8 or 6, along with 7 or 9 respectively. However, both 7 and 9 have been used. This is impossible. In the latter instance, the long side containing 4 must contain 1 and two other numbers with sum 15. Since exactly one of them is 7 or 8, this is also impossible.

Case 12.
The short side containing 4 consists of either 4+8+9 or 4+7+10. In the former instance, the long side containing 2 contains 9 and two other numbers with sum 10. Hence the number at the other corner cannot be 10. However, if it is 6, another copy of 4 is needed. This is impossible. In the latter instance, the other short side contains 6, 9 and another copy of 6, which is also impossible.

Case 13.
The short side containing 4 contains two other numbers with sum 17. Since 7 and 10 cannot both appear here, both belong to the other short side. However, this requires another copy of 4. Hence this case is impossible.

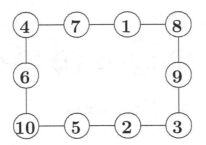

Figure 7.17

Puzzle 5.
When we compute the total of the six sums, each number is counted three times. Hence each sum is $3 \times (1+2+\cdots+8) \div 6 = 18$. If we let the numbers on the four slanting edges be (1,8), (2,7), (3,6) and (4,5) respectively, this will take care of the four lateral quadrilaterals. It is now a simple matter to adjust the numbers so that the inner and outer quadrilaterals are also taken care of. Figure 7.18 shows one of many possible placements of the numbers.

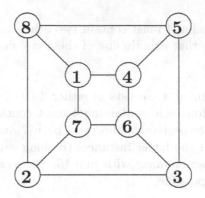

Figure 7.18

Puzzle 6.

We have 4=1+3, 9=1+8=2+7=3+6=4+5, 25=10+15=11+14=12+13 and 16=1+15=2+14=3+13=4+12=5+11=6+10=7+9. Every number appears twice on the right sides of these equations with four exceptions. Each of 1 and 3 appears three times, once together, while each of 8 and 9 appears only once. Thus we can line them up from 8 to 9 following the pairings on the right sides of the equations, omitting 4=1+3. The final configuration is shown in Figure 7.19.

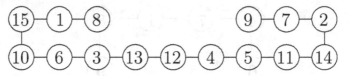

Figure 7.19

Puzzle 7.

(a) The triangles in Figure 7.20 are not hard to find. There are no other possibilities apart from reflections across the vertical axis.

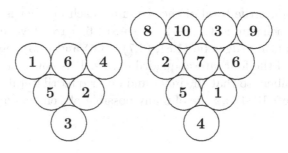

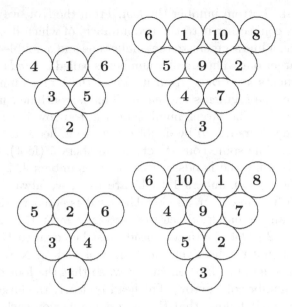

Figure 7.20

(b) The unique construction is shown in Figure 7.1. We now explain how it is found and why it is unique.

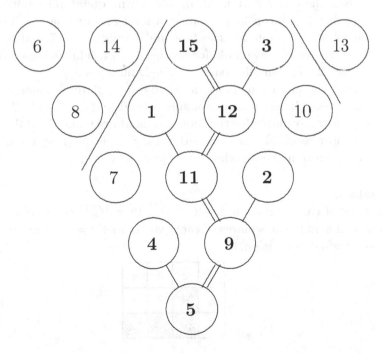

Figure 7.21

We call the bottom number the *foot*. From the foot, we draw two line segments connecting it to the two numbers of which it is the absolute difference, a line segment to the smaller one and a doubled line segment to the larger one. The smaller number is called a *hand* and no further line segments are drawn from it. From the larger number, we draw two more line segments as before. This is continued until we reach the top row. The larger number on the top row is called the *head*, and the numbers linked by doubled lines from head to foot constitute the *spine*. The spine consists of the numbers 5 (foot), 9, 11, 12 and 15 (head) while the hands consist of the numbers 4, 2, 1 and 3. No matter how the numbers 1 to 15 are used, we always have a spine of length five, and four hands. The head is equal to the sum of the foot and all the hands. Since the head is at most 15 while the sum is at least 1+2+3+4+5=15, the head must be equal to 15 and the foot and hands must collectively be 1, 2, 3, 4 and 5. Note that there is exactly one hand or foot on each row, so that the foot or hand is the smallest number of the row. The head is clearly the largest number of the top row. It follows that the largest number on each row is on the spine. From the head and the top hand, if we draw two slanting lines towards the sides, as shown in the diagram above, we can divide the whole structure into three parts. We have two cases.

Case 1. The smaller part is in fact empty.

Then the other part is an upside down equilateral triangle with six numbers. It has a foot, a spine, a head and two hands. The smallest numbers available for the foot and hands are 6, 7 and 8, while the largest number available for the head is 14. This is impossible.

Case 2. The smaller part is a single number.

Then the other part is an upside down equilateral triangle with three numbers. It has a foot, a spine, a head and one hand. The smallest numbers available for the foot and hand are 6 and 7, while the largest number available for the head is 14. Simple checking reveals that the construction given earlier is unique.

Puzzle 8.

The total of the numbers is $1 + 2 + \cdots + 12 = \frac{12 \times 13}{2} = 78$. Hence the total of the numbers in the squares of each part is $78 \div 2 = 39$. Figure 7.22 shows the only possible division.

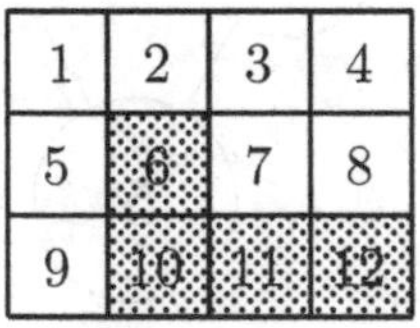

1	2	3	4
5	6	7	8
9	10	11	12

Figure 7.22

Puzzle 9.

The total of the numbers is $1 + 2 + \cdots + 16 = \frac{16 \times 17}{2} = 136$. Hence the total of the numbers in the squares of each part is $136 \div 2 = 68$. Let us focus on the part containing the number 16. If it also contains 13, 14 and 15, then we are only 10 short of 68. The only way to expand to 68 is to include 10 itself, as shown in Figure 7.23 on the left. Suppose it contains 14 and 15 but not 13. Then we are 23 short. We can include 11 and 12 as shown in Figure 7.23 in the middle, or include 6, 7 and 10 as shown in Figure 7.23 on the right.

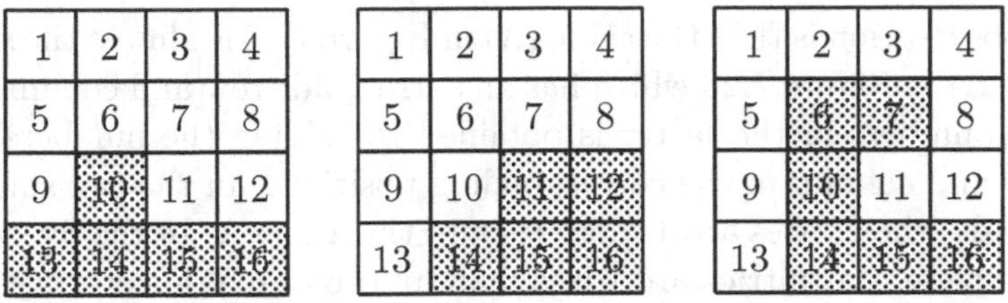

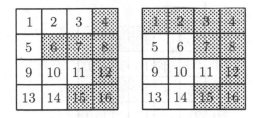

Figure 7.23

Suppose the part which contains 16 also contains 15 but not 14. If it also contains 11 and 12, then the total is only 14 short of 68. However, the 14 cannot be picked up without becoming disconnected or making the other part disconnected. Hence it must contain 12 but not 11. There are only two such divisions, shown in Figure 7.24.

Figure 7.24

Finally, suppose the part which contains 16 does not contain 15. Then it cannot contain 13 or 14 either. Since 13+14+15=42, the other part is only 26 short of 68. It cannot contain 10 since it is not possible to pick up another 16 without becoming disconnected or making the other part disconnected. Hence it must contain 9 and possibly 11. There are only two such divisions, shown in Figure 7.25.

1	2	3	4
5	6	7	8
9	10	11	12
13	14	15	16

1	2	3	4
5	6	7	8
9	10	11	12
13	14	15	16

Figure 7.25

Puzzle 10.

The key observation is that the table (A) in Figure 7.10 is almost the same as the table (B) in Figure 7.26 which has an extra guide row and column. Each of the 25 numbers in the latter is obtained by adding the numbers in the guide row and column in the corresponding positions. In the two squares in a light shade, the entries are 1 more than in the Table A. In the five squares in a deep shade, the entries are 1 less than in Table A. Whenever we choose 5 squares from Table B, with one in each row and one in each column, the sum of the five numbers in them is always 51, which is the sum of the five numbers in the guide row and the five numbers in the guide column. So we should avoid the lightly shaded squares and choose as many of the darkly shaded squares as possible. We can only choose two of them since they are all in two columns, so that the highest sum we can reach is 51+2=53. In Table A, we must take the 13 from the third column and one of 13, 8 and 14 the fourth. For instance, we can take 21, 6, 13, 13 and 0 from the five columns left to right.

+	11	2	7	3	0
5	16	7	12	8	5
9	20	11	16	12	9
4	15	6	11	7	4
10	21	12	17	13	10
0	11	2	7	3	0

Figure 7.26

Puzzle 11.

This is a fair puzzle. The special property in the mind of the author is that the sum of any four numbers in a 2×2 subtable is always 20. By checking the three numbers in such a subtable other than the number covered up, we can tell what that number must be. There may be other properties that also work, but we only need one.

Puzzle 12.

We cannot touch 5 or 7, but we can make use of the following equations: $1 \times 4 = 2 \times 2$, $1 \times 6 = 2 \times 3$, $1 \times 8 = 2 \times 4$, $1 \times 9 = 3 \times 3$, $2 \times 6 = 3 \times 4$, $2 \times 8 = 4 \times 4$, $2 \times 9 = 3 \times 6$, $3 \times 8 = 4 \times 6$ and $4 \times 9 = 6 \times 6$. Suppose we replace 3 and 6 with 2 and 9. We still have nine numbers with the correct product, but the sum has increased by 2. This can be fixed if we now replace 2 and 8 with 4 and 4. So 1, 2, 4, 4, 4, 5, 7, 9 and 9 are another possibility.

Puzzle 13.

Since 18+13+14=45, all prizes were claimed, and the three targets in the top row were all hit. Since Colin shot last and he could not get 14 from only one string, both of his shots hit targets in the top row. If he hit the target above the 4, he could not also have got the 7. This is because he needed the 3 to make 14, but the target above it was not in the top row. If he got only the 4, he could not have got 10 from his other shot. Hence he did not hit the target above the 4, so that he must have hit the other two targets in the top row. The only way for him to get 14 was to claim 5+1 from the string in the middle and 8 from the string on the right. To get 13 in total, Brian must claim all three prizes from the string on the left. Hence he hit the target in the top row after he had hit one of the other two targets below. Finally, Alice must have hit the targets above the 6 and the 9.

Puzzle 14.

We must hit 37 at least once. Otherwise we cannot get more than $4 \times 23 = 92$ points. The unit digits of the scores are 1, 3, 7 and 9, and the sum of four of them must be a multiple of 10. The possible combinations involving at least one 7 are 1+1+1+7, 1+3+7+9, 3+3+7+7 and 7+7+7+9. The respective sums so far are 10, 20, 20 and 30. In the first case, we must add 10+10+10+30 and get 70. In the second case, we must add 10+20+30+10 and get 90. In the third case, we may add 20+20+30+10 and get 100. In the fourth case, if we only use one copy of 37, we get 90. If we use at least two copies of 37, we get at least 110. Hence the only way to get 100 is hit 23 twice and each of 17 and 37 once.

Puzzle 15.

The other player who uses locker room number 10 is player number 3. Since there are 5 other even-numbered players and only 3 other odd-numbered players, some locker room is being used by two even-numbered players, and this must be locker room number 12. Suppose the players are number 8 and number 4. Now 13=10+3=9+4=8+5=7+6, but one player from each sum is using a locker room other than 13. This contradiction shows that the players who use locker room number 12 are number 10 and number 2. Now player number 1 can only share locker room number 9 with player number 8. The other combinations are players number 6 and number 5 in locker room number 11, and players number 9 and number 4 in locker room number 13.

Puzzle 16.

Since 10+5+7=22 while 1+2+3+4+5+6=21, the six players in the numbered locker rooms must be numbers 1, 2, 3, 4, 5 and 7. Player number 7 must be in locker room number 10, whose other occupant must be player number 3. Player number 5 must be in locker room number 7, whose other occupant must be player number 2. It follows that players number 1 and number 4 must be in locker room number 5.

Puzzle 17.

The cheetah takes 34 hops to reach the end of the track, ending up 2 meters beyond it. It will take another 34 hops to get back, and the total time is 34 minutes. The fox takes 50 hops each way, and the total time is 33 minutes and 20 seconds. Thus the fox wins the race.

Puzzle 18.

There are 14 minutes between 4:56 pm and 5:10 pm, and 14=4+10. Hence the correct time is between 4:56 pm and 5:10 pm. Now the second clock is off by 5 minutes. Hence the correct time is either 4:56 pm or 5:06 pm. The former is impossible as otherwise the first clock is correct. It follows that the latter is the case, and the correct time is 5:06 pm.

Puzzle 19.

Suppose the soldiers without uniform form an $x \times y$ array with $x \geq y$. Then all the soldiers form an $(x+2) \times (y+2)$ array, and we have $(x+2)(y+2) = 2xy$. This may be rewritten as $4 = xy - 2x - 2y$, so that $8 = xy - 2x - 2y + 4$, which may be factorized as $(x-2)(y-2)$. Since x and y are positive integers, we have either $x - 2 = 8$ and $y - 2 = 1$, or $x - 2 = 4$ and $y - 2 = 2$. So the inner array is either 10×3 or 6×4, and the outer array is either 12×5 or 8×6. It follows that the larger unit had 60 soldiers and the smaller one 48 soldiers.

Puzzle 20.

Clearly, the number of pennies must be a multiple of 10. If all 50 coins are pennies, the total value is only 50 cents. If at most 30 of them is a penny, then there are at least 20 coins which is worth at least 5 cents each, so that the total value will exceed 1 dollar. It follows that the number of pennies is exactly 40, contributing 40 cents to the total. The remaining 60 cents must be made up with 10 coins, each either a nickel or a dime. If they are all nickels, the total is only 50 cents. Trading 1 nickel for 1 dime raises the total by 5 cents. Hence we have to trade 2 nickels for 2 dimes. In summary, there are 40 pennies, 8 nickels and 2 dimes.

Puzzle 21.

At most one of the bills is of 50 Euros. We consider two cases.

Case 1. A bill of 50 Euros is used.

The other 20 bills must make up the remaining amount of 50 Euros. If all bills are of 1 Euro, the total is way too low. If no bills are of 1 Euro, then the total of these 20 bills is at least 100 Euros, which is 50 Euros too much. Hence bills of 1 Euro must be used, and they must come in sets of 5. Start with 20 bills of 5 Euros. If we trade in one set of bills of 1 Euro for one set of bills of 5 Euros, we reduce the total by 20 Euros. Trading in two sets of 5 bills is still not enough, but as noted before, we cannot trade in four sets. Suppose we trade in three sets. The total is now down to 40 Euros, which is 10 Euros too low. To recover the deficit, we can trade in 2 bills of 5 Euros for 2 bills of 10 Euros. Thus there is a unique solution in this case, consisting of 1 bill of 50 Euros, 2 bills of 10 Euros, 3 bills of 5 Euros and 15 bills of 1 Euro.

Case 2. No bills of 50 Euros are used.

If no bills are of 1 Euro, then the total of the 21 bills is at least 105 Euros. Hence bills of 1 Euro must be used, in sets of 5. If 3 sets are used, then even if the other 6 bills are all of 10 Euros, the total is only 75 Euros. Hence we must use 1 or 2 sets. We consider these subcases separately.

Subcase 2(a). Only one set of 5 bills of 1 Euro is used.

The remaining 16 bills are either of 5 Euros or of 10 Euros, and are worth 95 Euros in total. If they are all of 5 Euros, we will be 15 Euros short. So we must trade in 3 bills of 5 Euros for 3 bills of 10 Euros. The unique solution in this subcase consists of 3 bills of 10 Euros, 13 bills of 5 Euros and 5 bills of 1 Euro.

Subcase 2(b). Two sets of 5 bills of 1 Euro are used.

The remaining 11 bills are either of 5 Euros or of 10 Euros, and are worth 90 Euros in total. If they are all of 5 Euros, we will be 35 Euros short. So we must trade in 7 bills of 5 Euros for 7 bills of 10 Euros. The unique solution in this subcase consists of 7 bills of 10 Euros, 4 bills of 5 Euros and 10 bills of 1 Euro.

In summary, the task can be accomplished in three different ways.

Puzzle 22.

Apart from $29 and $70, all the other prices are multiples of $3. Now $100 is not a multiple of $3 either. The remainder when 29, 70 and 100 are divided by 3 are 2, 1 and 1 respectively. Hence volume 9, which costs $29, cannot have been sold while volume 12, which costs $70, must have been sold. It is easy to see that the other $30 must have come from the sale of volumes 3 and 7, which cost respectively $18 and $12.

Puzzle 23.

Consider a highway of length 60 meters. In the first plan, there are 3 trees to be planted. The total cost is $600, so that the cost is $200 per tree. In the second plan, there are 6 trees to be planted. So the cost is $1,200, which agrees with the cost per tree computed above. In the third plan, there are 4 trees to be planted. Hence the cost is $200 × 4=$800. The actual highway is most certainly longer than 60 meters, but by scaling the above calculations, we obtain the same answer.

Puzzle 24.

The total pay is $270 and the second worker has done $\frac{5}{9}$ of the work. So his total pay should be $150, and he should get $60 of the $90 that is going to the first worker. The third worker gets the other $30.

Puzzle 25.

The total amount of gold is 4+9+11+12+19=55 kilograms. The total amount of gold received by the son and his grandmother is a multiple of 3 kilograms. Divided by 3, the respective remainders from 4, 9, 11, 12 and 19 are 1, 0, 2, 0 and 1. This means that the wife has exactly one of the 4-kilogram and the 19-kilogram gold bars but does not have the 11-kilogram gold bar. The following chart summarizes the eight cases.

Wife has	Others have	Mother Has	Status
4	55 − 4 = 51	17	Impossible
4+9=13	55 − 13 = 42	14	Impossible
4+12=16	55 − 16 = 39	13=4+9	Impossible
4+9+12=25	55 − 25 = 30	10	Impossible
19	55 − 19 = 36	**12**	**Possible**
19+9=28	55 − 28 = 27	9	Impossible
19+12=31	55 − 31 = 24	8	Impossible
19+9+12=40	55 − 40 = 15	5	Impossible

There is only one possible case, where the wife has the 19-kilogram gold bar, the mother has the 12-kilogram gold bar and the son has the other three.

Chapter Eight
Geometric Puzzles

Puzzle 1.

$ABCD$ is a long rectangular piece of paper. When folded along EF as shown in Figure 8.1, with AB landing on $A'B'$, the distance between $A'B'$ and CD is 2 centimeters.

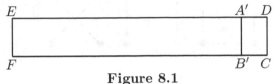

Figure 8.1

When folded along GH as shown in Figure 8.2, with CD landing on $C'D'$, the distance between $C'D'$ and AB is also 2 centimeters.

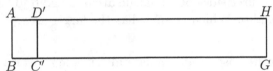

Figure 8.2

What is the distance between EF and GH?

Puzzle 2.

An old ruler is 13 centimeters in length. The markings are fading, and only those at 1 centimeter, 2 centimeters and 6 centimeters from one end are visible. Thus we can use this rule to measure exact lengths of 1, 1+1=2, 4, 4+1=5, 4+1+1=6, 7, 7+4=11, 7+4+1=12 and 7+4+1+1=13 centimeters. Where should a fourth marking be added so that we can also measure the remaining lengths of 3, 8, 9 and 10 centimeters?

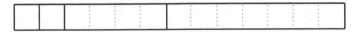

Figure 8.3

Puzzle 3.

Four communities are located at the corners of a square of side length 100 kilometers. Anyone from any of the towns should be able to call anyone else in the same town or any other town, relayed at most twice. The telephone company charges the towns by the total length of telephone wires. The initial plan has them running along all four sides of the square, for a total length of 400 kilometers.

99

A. Liu et al., *The Puzzles of Nobuyuki Yoshigahara*, Problem Books in Mathematics, https://doi.org/10.1007/978-3-030-62896-3_8

Later, at some inconvenience to the people of two of the towns, one of the four segments of telephone wires was removed to reduce the total length to 300 kilometers. Then an improved plan has the telephone wires run along both diagonals of the square, with a telephone exchange at their intersection. This improves service and actually reduces the cost since the total length is now $2 \times 100\sqrt{2}$ or just under 283 kilometers. Is further improvement possible?

Figure 8.4

Puzzle 4.
The lengths of the three sides of a triangle are 5, 5 and 6. Those of another triangle are 5, 5 and 8. Which triangle has the larger area?

Puzzle 5.
A line parallel to the base of an equilateral triangle is drawn at a distance above the base equal to one third of the distance of the vertex above the base, as shown in Figure 8.5. What is the ratio of the area of the top part to the area of the bottom part?

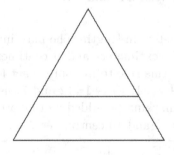

Figure 8.5

Puzzle 6.
An equilateral triangle of side length 1 is divided by a continuous curve into two parts of equal area. What is the minimum length of this curve?

Puzzle 7.
A triangle is divided into three triangles and a quadrilateral by two lines, as shown in Figure 8.6, where the areas of the triangles are given. What is the area of the quadrilateral?

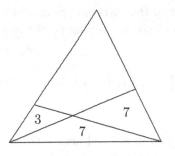

Figure 8.6

Puzzle 8.
Draw seven lines on the plane and try to enclose as many triangles as possible. Any two of these triangles may share at most one common point.

Puzzle 9.
Arrange some unit circles on the plane so that each is tangent to exactly three of the others.

Puzzle 10.
Figure 8.7 shows that seven unit circles can fit inside a circle of radius 3. How many unit circles can fit inside a circle of radius 4?

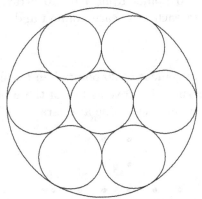

Figure 8.7

Puzzle 11.
Four points on the plane determine six pairwise distances. Arrange the points so that there are only two distinct values among these six distances. Both values must be positive.

Puzzle 12.

(a) Remove all but five of the twenty-five points in a 5 × 5 array so that the pairwise distances of the remaining points are distinct.

(b) Remove all but six of the thirty-six points in a 6×6 array so that the pairwise distances of the remaining points are distinct.

Puzzle 13.
Place 9 points on the plane so that each is at the same distance from four of the other points.

Puzzle 14.
Sixteen points are arranged in a 4×4 array. Draw a closed loop consisting of six line segments joined end to end, such that the loop passes through each point at least once.

Puzzle 15.
Sixteen points are arranged in a 4×4 array. Remove six of them so that the number of rows, columns and diagonals containing an even number of the remaining points is

(a) as large as possible;

(b) as small as possible.

Puzzle 16.
Is it possible to remove 6 points from a 6×6 array, leaving behind an even number of points in each row, each column and each of the two long diagonals?

Puzzle 17.
Figure 8.7 shows an arrangement of 20 points in a 6×6 array with a 2×2 array missing at each corner. Remove as few of these points as possible so that no four remaining points determine a square.

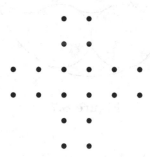

Figure 8.7

Puzzle 18.
Figure 8.8 shows a 3 × 3 square with the central unit square removed, and an additional cut is made between two of the unit squares so that it becomes a crooked paper strip. Use it to cover up the surface of a unit cube.

Figure 8.8

Puzzle 19.
Figure 8.9 shows three 3 × 3 cards with five or six holes in each. The cards may be rotated or reflected. Put them in a stack and minimize the number of holes which pass through the entire stack. What is the smallest number of such holes?

Figure 8.9

Puzzle 20.
The Eiffel Tower is 324 meters high and weighs 7301384 kilograms. The Eyeful Tower, an exact replica, is to be constructed with the same material but only to a height of 162 meters. The project manager sends the government a bill for 7301384 ÷ 2 = 3650692 kilograms of material, and is charged with fraud. Why?

Puzzle 21.
Twenty-three rectangular blocks with uniform thickness are placed on a flat surface. Figure 8.10 shows the view of them from the top. All but one block are lying horizontally. Which block is not?

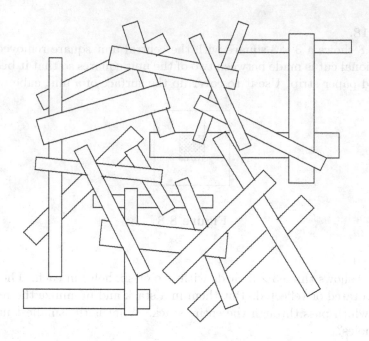

Figure 8.10

Puzzle 22.

Several identical coins form a structure on a flat surface. The top view is shown in Figure 8.11 on the left. The side view of the structure in the direction indicated by the arrow is shown in Figure 8.11 on the right. How many coins are in the structure?

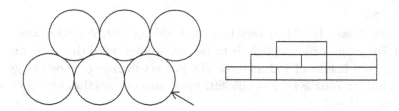

Figure 8.11

Puzzle 23.

Figure 8.12 shows a white block with an indentation interlocking with a shaded block with a protrusion. How can they be put together?

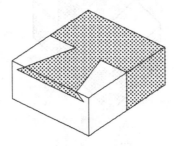

Figure 8.12

Puzzle 24.

Each of the pieces in Figure 8.13 consists of four unit cubes.

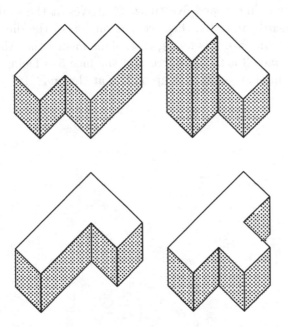

Figure 8.13

Use two different pieces to form the shape in Figure 8.14. The pieces may be rotated, and the shading may be ignored.

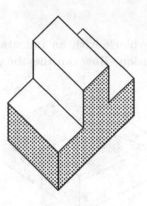

Figure 8.14

Puzzle 25.

A cubical die is sitting on the northwest corner square of a 4 × 3 board. Each square of the board is the same size as a face of the die. The die is then tipped either to the east or to the south, provided that it remains on the board. After being tipped five times, it arrives at the southeast corner square of the board. Initially, the face 1 is on top of the die, the face 4 is facing south and the face 2 is facing east. This means that the face 6 is at the bottom, the face 3 is facing north and the face 5 is facing west. Which of these faces cannot possibly end up on top at the end?

Solutions

Puzzle 1.

Figure 8.15 shows the bottom edge of the rectangles.

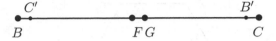

Figure 8.15

Note that $BF = B'F$ and $CG = C'G$. We also have $B'C = C'B = 2$ centimeters, so that $BG = CF$ and $B'F = C'G$. It follows that we have $FG = BG - BF = BC' + C'G - B'F = 2$ centimeters.

Puzzle 2.

In order to be able to measure a length of 10 centimeters, the fourth marking must be at 3, 10, 11 or 12 centimeters from the same end, as shown in Figure 8.16.

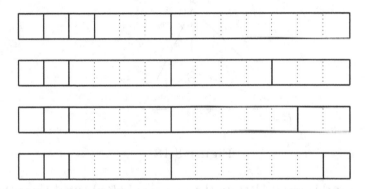

Figure 8.16

It is possible to measure a length of 8 centimeters only in the second case. All four missing distances, 3, 4+4=8, 4+4+1=9 and 4+4+1+1=10, can be measured here.

Puzzle 3.

Further improvement is indeed possible. Figure 8.17 shows the new plan with two telephone exchanges. All telephone wires that come together make angles of 120°. The total length is only $4 \times 25\sqrt{3} + (100 - 25\sqrt{3})$ or just under 230 kilometers.

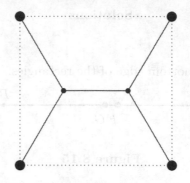

Figure 8.17

Further improvement is now impossible. A point where three line segments meet to form three 120° angles is called a *Fermat* point. Every triangle ABC contains a Fermat point F provided that its largest angle is less than 120°. For any point P inside ABC, $PA + PB + PC \geq FA + FB + FC$, with equality if and only if $P = F$.

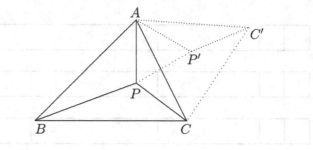

Figure 8.18

We now prove this result. Rotate the triangle CAP about A through a 60° angle to its new position $C'AP'$, as shown in Figure 8.18. Then PAP' is an equilateral triangle, $P'C' = PC$ and $\angle AP'C' = \angle APC$. Now $PA + PB + PC = PP' + BP + P'C' \geq BC'$, with equality if and only if B, P, P' and C' are collinear. Note that this happens if and only if $\angle BPA = 120° = \angle AP'C'$, or equivalently, P is a Fermat point.

Puzzle 4.

As shown in Figure 8.19, both triangles are composed of two right triangles with side lengths 3, 4 and 5. Hence they have the same area.

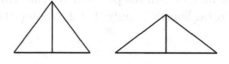

Figure 8.19

Puzzle 5.

The top part is also an equilateral triangle. The side lengths of the two triangles are in the ratio of 3:2. Hence their areas are in the ratio of 9:4 so that the desired ratio is 4:5. This answer also emerges graphically if we draw additional lines as shown in Figure 8.20.

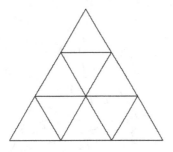

Figure 8.20

Puzzle 6.

Most people choose between a median of the triangle or a straight line parallel to a side of the triangle. If we put six copies of the triangle together at a common vertex, as shown in Figure 8.21, the dividing segments link up to enclose a shaded region which has half the area of the hexagon. Since the hexagon has a fixed area, this region also has a fixed area. A well-known result called the Isoperimetric Theorem states that of all curves which enclose a region of fixed area, the circle is the shortest. It follows that the minimal dividing line is one-sixth of a circle of radius approximately 0.64, so that the length of the dividing arc is approximately 0.67. For comparison, the length of the median is approximately 0.87, and the length of the dividing segment parallel to a side of the triangle is approximately 0.71.

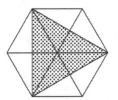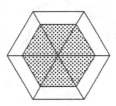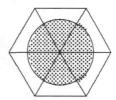

Figure 8.21

Puzzle 7.
We divide the quadrilateral into two triangles of respective area x and y, as shown in Figure 8.22. Then $\frac{x}{3} = \frac{y+7}{7}$ and $\frac{y}{7} = \frac{x+3}{7}$. These simplify to $7x = 3y + 21$ and $y = x + 3$. Hence $7x = 3(x+3) + 21$ so that $4x = 30$. It follows that $x = 7.5$ and $y = x + 3 = 10.5$. The area of the quadrilateral is 7.5+10.5=18.

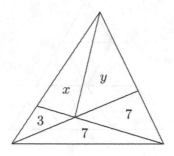

Figure 8.22

Puzzle 8.
Figure 8.23 shows that as many as eleven triangles may be enclosed.

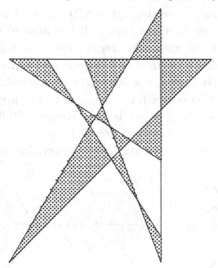

Figure 8.23

Puzzle 9.
We can form a group of four unit circles so that each of two of them is tangent to all the others and each of the other two is tangent to only two others. The condition of the puzzle can be met if we put together four such groups. Sixteen unit circles is believed to be minimum.

110

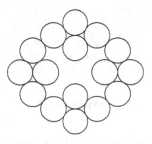

Figure 8.24

Puzzle 10.
Figure 8.25 shows that eleven unit circles can fit inside a circle of radius 4.

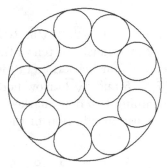

Figure 8.25

Puzzle 11.
The six distances may be divided into two classes 5:1, 4:2 or 3:3. In the first case, we must have two equilateral triangles with a common side, as shown in Figure 8.26 on the left. In the second case, we may have three of the four equal distances involving the same point. Then the fourth one completes an equilateral triangle. There are two possibilities, as shown in Figure 8.26 in the middle and on the right.

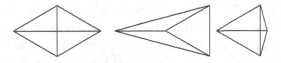

Figure 8.26

If no three of the four equal distances involve the same point, then these distances define a rhombus. Since the other two distances are equal, the rhombus is in fact a square, as shown in Figure 8.27 on the left.

111

Figure 8.27

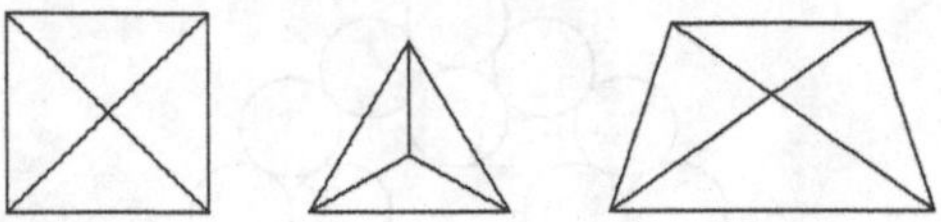

In the third case, one set of three equal distances may define an equilateral triangle. Then the fourth point must be the center of this triangle, as shown in the diagram above in the middle. If neither set defines an equilateral triangle, each set is a path joined end to end, with a self-intersection in one of them. This is shown in Figure 8.27 on the right.

Puzzle 12.

(a) The available distances are 1, $\sqrt{2}$, 2, $\sqrt{5}$, $2\sqrt{2}$, 3, $\sqrt{10}$, $\sqrt{13}$, 4, $\sqrt{17}$, $3\sqrt{2}$, $2\sqrt{5}$, 5 and $4\sqrt{2}$. We must use ten of these fourteen as pairwise distances. If we keep the four black dots in Figure 8.28 on the left, we have used 1, 3, $\sqrt{17}$, 4, 5 and $4\sqrt{2}$. Now none of the white dots can be kept. Of the remaining two dots marked by targets, any of them may be kept, making use in addition of the distances $\sqrt{2}$, $\sqrt{5}$, $\sqrt{10}$ and $\sqrt{13}$.

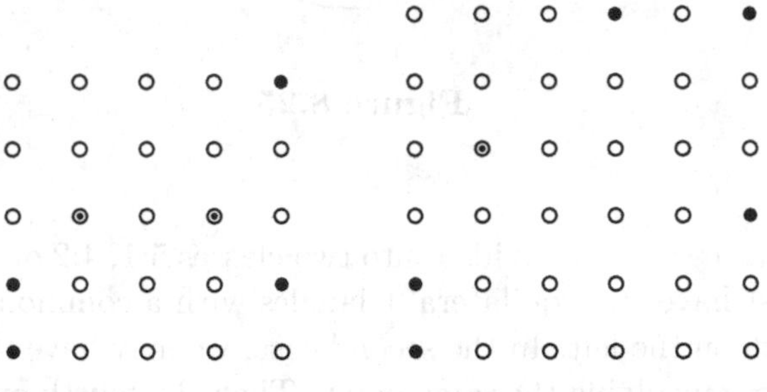

Figure 8.28

(b) In addition to the distances in (a), available also are $\sqrt{26}$, $\sqrt{29}$, $\sqrt{34}$, $\sqrt{41}$ and $5\sqrt{2}$. We must use fifteen of these nineteen as pairwise distances. If we keep the five black dots in Figure 8.28 on the right, we can only add the dot marked by a target. This configuration uses the distances 1, 2, $\sqrt{5}$, $2\sqrt{2}$, 3, $\sqrt{10}$, $\sqrt{13}$, $\sqrt{17}$, $2\sqrt{5}$, 5, $\sqrt{26}$, $\sqrt{29}$, $\sqrt{34}$, $\sqrt{41}$ and $5\sqrt{2}$.

Puzzle 13.
Figure 8.29 shows a placement utilizing three squares and six equilateral triangles. It should be pointed out that the structure is not rigid. Slight displacement yields infinitely many other solutions.

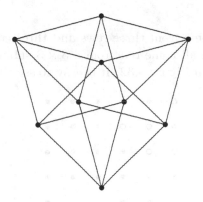

Figure 8.29

Puzzle 14.

Six segments are insufficient if all of them are either horizontal or vertical. With slanted segments extending beyond the 4 × 4 frame, this is then sufficient. Figure 8.30 shows two of many such loops.

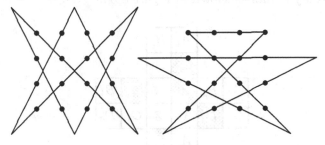

Figure 8.30

Puzzle 15.

(a) Figure 8.31 on the left shows that we can have as many as 14 such lines.

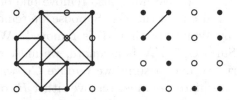

Figure 8.31

(b) Figure 8.31 on the right shows that we may have as little as 1 such line.

113

Puzzle 16.

The 6 points must come from three rows and three columns, with 2 from each. Hence they can all come out from a 3×3 subboard, say at the top left corner, as shown in Figure 8.32. It is easy to see that all conditions are satisfied.

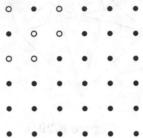

Figure 8.32

Puzzle 17.

Label the twenty points as shown in Figure 8.33.

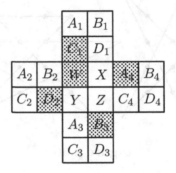

Figure 8.33

Clearly, we must remove one of W, X, Y and Z, as well as one of A_k, B_k, C_k and D_k for $k = 1, 2, 3, 4$. We must also remove one of the As, one of the Bs, one of the Cs and one of the Ds. Suppose we remove only five points. We may assume by symmetry that W is removed. We shade this in the diagram above. Since $C_1 B_2 Y X$ is a square, we must remove either C_1 or B_2. By symmetry, we may assume we remove C_1, which is then shaded. Since XYB_3C_4 is a square, we must remove either B_3 or C_4. Since we have already removed C_1, we must remove B_3. Similarly, we must remove A_4 since XZC_4A_4 is a square. The last square to be removed must therefore be D_2. However, now both of the squares $A_1C_2D_3B_4$ and $B_1A_2C_3D_4$ remain intact. It follows that we must remove at least six dots. This can be accomplished by removing exactly six points, as shown in Figure 8.34.

114

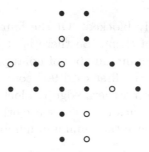

Figure 8.34

Puzzle 18.
Of the 35 hexominoes, figures consisting of 6 unit squares joined edge to edge, only 11 can form the net of a cube. These are shown in Figure 8.35.

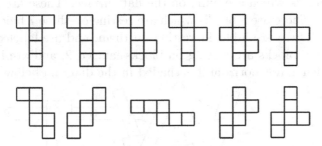

Figure 8.35

Label the eight squares of our paper strip as shown in Figure 8.36 on the left. After the first fold, the paper strip takes on the shape shown in Figure 8.36 in the middle. After the second fold, it takes on the shape in Figure 8.36 on the right.

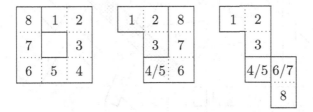

Figure 8.36

We now have the first hexomino in the second row. If we let the bottom face of the unit cube rest on square 3, then square 2 covers the back face and square 1 covers the left face. Square 4/5 covers the front face, square 6/7 covers the left face and square 8 covers the top face.

115

Puzzle 19.

The central space is clearly blocked. Of the four spaces along the edges, each card can block one of them. So there is at least one space which is unblocked. Hence the minimum number of holes passing through any stack is at least 1. If we rotate the first card 90° counterclockwise, every space except the space along the northern edge is blocked. Thus the number of holes passing through this stack is 1. There are other ways of stacking these three cards which also achieve the minimum number of holes.

Puzzle 20.

The height of the Eyeful Tower is $\frac{1}{2}$ that of the Eiffel Tower. Hence its volume is $(\frac{1}{2})^3 = \frac{1}{8}$ that of the Eiffel Tower. Since weight is proportional to volume, the amount of material required is only $\frac{1}{8} \times 7301384 = 912673$ kilograms.

Puzzle 21.

There are 7 blocks which are lying on the flat surface. These are labeled 0 in Figure 8.37. There are 8 blocks which are resting on these 7 blocks. They are labeled 1. There are 5 blocks resting on them, and are labeled 2. Two of the remaining blocks are resting on blocks labeled 2, and are labeled 3. The block which is not horizontal is shaded in the diagram below.

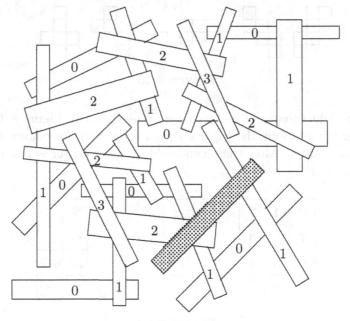

Figure 8.37

Puzzle 22.

Label the six stacks of coins A, B, C, D, E and F as shown in Figure 8.38 on the left.

116

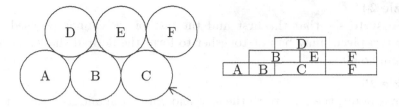

Figure 8.38

For each totally or partially visible coin in the side view, the stack to which it belongs can be determined, as shown in Figure 8.28 on the right. There is only one coin in stack A. There are only two coins in stack B as otherwise we could not have seen totally the coin in stack D. There is only one coin in stack C as otherwise we could not have seen totally the coin in stack B. There are three coins in stack D. There are only two coins in stack E as otherwise we could not have seen totally the coin in stack D. Finally, there are two coins in stack F. The total is 1+2+1+3+2+2=11.

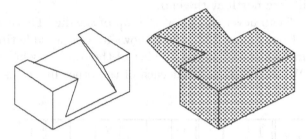

Figure 8.39

Puzzle 23.
Neither the indentation nor the protrusion is flat, but is wedge-shaped, as shown in Figure 8.39. The shaded piece starts lower than the white piece and is then pushed up into it.

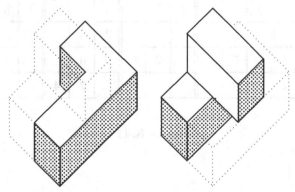

Figure 8.40

Puzzle 24.

It is easy to see that the first and the last pieces cannot be used. The other two pieces may be put together to form the desired shape, as shown in Figure 8.40.

Puzzle 25.

At some point, the die tips to the east and enters the middle column for the first time. Since the face 5 is initially facing west, it will now be on top. We consider four cases.

Case 1. The tipping continues eastward.

The face 5 will then face east and remains facing east to the end.

Case 2. The tipping goes one square south before turning east.

The face 5 will then face south. The die now tips to the south till the end, at most twice. This is not enough to bring the face 5 to the top.

Case 3. The tipping goes two squares south before turning east.

The face 5 will be at the bottom before the turning, and face west after. It remains facing west.

Case 4. The tipping goes three squares south before turning east.

The face 5 will face north at the end.

Hence the face 5 can never end up at the top of the die. This conclusion can also be drawn from Figure 8.41 which shows all ten possible tipping routes. The number in each square is the number of the top face when the die is on that square. We see moreover that each of the other five faces may end up on top.

1	5	6	1			1			1			1	5	
		3	3	5	4	3			3				3	6
		1			6	6	5	1	6					2
		4			**3**			**4**	4	5	**3**			**1**

1			1			1	5		1			1	5	
3	5		3				3		3	5			3	
	6	4	6	5			2	6		6			2	
		2		4	1			**4**		2	**4**		4	6

Figure 8.41

118

Chapter Nine
Dissection Puzzles

Puzzle 1.
Figure 9.1 shows an estate with 4 gardens, which are shaded, along with 12 buildings. Divide the estate along grid lines into four congruent parts, with a garden in each. Ignore the internal structure of the parts and the width of the lanes between the gardens and buildings. *We adopt the convention that two figures obtainable from each other by rotation or reflection are considered congruent.*

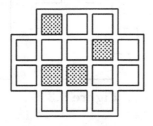

Figure 9.1

Puzzle 2.
Figure 9.2 shows an estate with 4 gardens, which are shaded, along with 19 buildings. Each garden or building takes up 1 or 2 squares. Divide the estate along grid lines into four congruent parts, with a garden in each. Each garden or building taking up 2 squares may not be divided between two parts. Ignore the internal structure of the parts and the width of the lanes between the gardens and buildings.

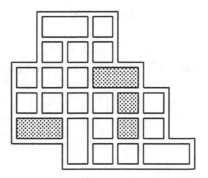

Figure 9.2

© The Author(s), under exclusive license to Springer Nature Switzerland AG 2020
A. Liu et al., *The Puzzles of Nobuyuki Yoshigahara*, Problem Books in Mathematics,
https://doi.org/10.1007/978-3-030-62896-3_9

Puzzle 3.
Dissect along grid lines five of the six shapes in Figure 9.3 into three congruent parts each. Which is the one for which this is impossible?

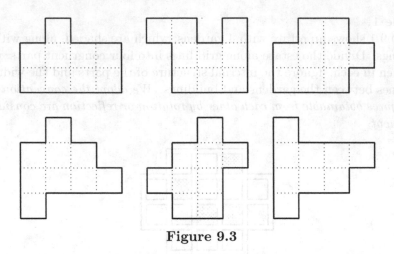

Figure 9.3

Puzzle 4.
Dissect the shape in Figure 9.4 into four congruent parts.

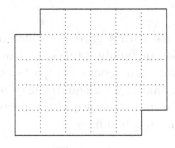

Figure 9.4

Puzzle 5.
Dissect the shape in Figure 9.5 into four congruent parts.

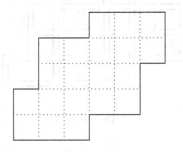

Figure 9.5

Puzzle 6.

Dissect the shape in Figure 9.6 into four congruent parts.

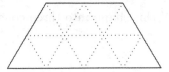

Figure 9.6

Puzzle 7.

Dissect the shape in Figure 9.7 into two congruent parts.

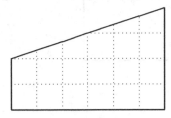

Figure 9.7

Puzzle 8.

Dissect the shape in Figure 9.8 into two congruent parts.

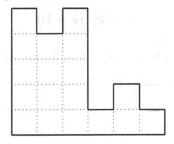

Figure 9.8

Puzzle 9.

The shape in Figure 9.9 has been dissected into four congruent parts by the dashed lines.

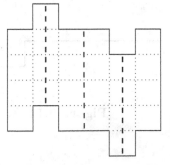

Figure 9.9

(a) Find two other dissections of this shape into four congruent parts by cutting along the grid lines.

(b) Find a dissection of this shape into three congruent parts.

Puzzle 10.

(a) Dissect the shape in Figure 9.10 into two congruent parts.

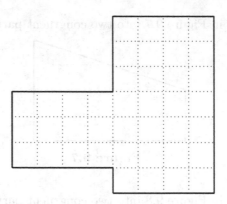

Figure 9.10

(b) Dissect the shape in Figure 9.11 into four congruent parts.

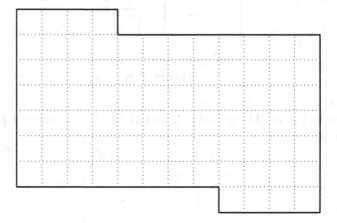

Figure 9.11

Puzzle 11.

A pentomino is a shape consisting of five unit squares joined edge to edge. The twelve pentominoes are shown in Figure 9.12.

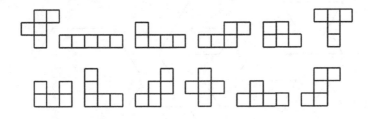

Figure 9.12

Dissect the shape in Figure 9.13 into two pentominoes.

Figure 9.13

Puzzle 12.

Dissect a square into several isosceles right triangles where no two triangles are of the same size.

Puzzle 13.

Dissect a rectangle into five smaller rectangles such that their five lengths and five widths are the numbers 1, 2, 3, 4, 5, 6, 7, 8, 9 and 10 in some order.

Puzzle 14.

Assemble four of the five pieces in Figure 9.14 on the left to form the shape in Figure 9.14 on the right.

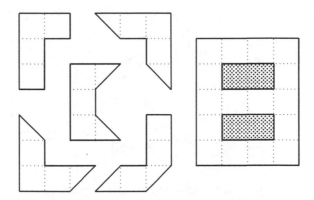

Figure 9.14

Puzzle 15.
Assemble four of the five pieces in Figure 9.15 on the left to form the regular hexagon in Figure 9.15 on the right.

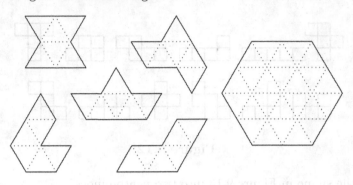

Figure 9.15

Puzzle 16.
We have seven copies of the piece shown in Figure 9.16 on the left, and two copies of the piece shown in Figure 9.16 on the right. Each piece consists of 4 unit regular hexagons joined edge to edge.

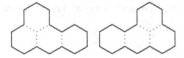

Figure 9.16

Assemble these nine pieces to form the shape in Figure 9.17. The pieces may *not* be reflected.

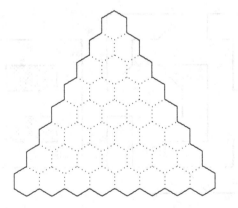

Figure 9.17

Puzzle 17.

$ABCD$ is a square. E is a point on CD such that $\angle DAE = 30°$, and F is a point on DA such that $\angle FED = 15°$. G is the foot of perpendicular from F to AE. $ABCD$ is cut along AE and FG into three pieces. Two more squares of the same size are cut in identical fashion. Assemble the nine pieces to form a single square.

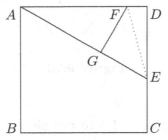

Figure 9.18

Puzzle 18.

Four teams take part in a week-long tournament in which every team plays every other team twice, and each team plays one game per day. Figure 9.19 on the left shows the final scoreboard, part of which has broken off into four pieces, as shown in Figure 9.19 on the right. These pieces are printed only on one side. A black circle indicates a victory and a white circle indicates a defeat. Which team wins the tournament?

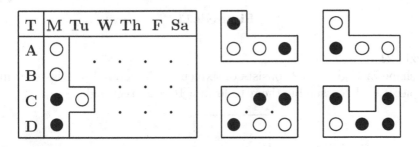

Figure 9.19

Puzzle 19.

Dissect the shape in Figure 9.20 into three pieces and reassemble them to form a rectangle. Four of the seven angles of the shape are right angles.

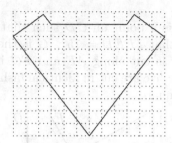

Figure 9.20

Puzzle 20.

Dissect the shape in Figure 9.21 into two pieces and reassemble them to form a 3×5 rectangle.

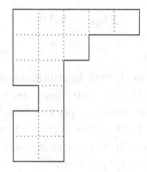

Figure 9.11

Puzzle 21.

The shape in Figure 9.22 consists of sixteen 4×5 rectangles. Dissect it into two pieces and reassemble them to form a 16×20 rectangle.

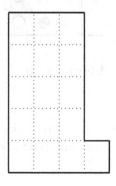

Figure 9.22

Puzzle 22.

Dissect one shape in Figure 9.23 into three pieces and reassemble them to form the other shape.

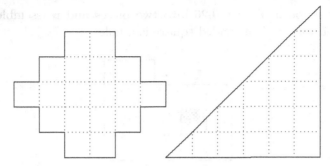

Figure 9.23

Puzzle 23.

Figure 9.24 on the left shows a unit square with one quadrant missing. The diagram below on the right shows a diamond shape of the same area. The shaded part is half a unit square and the three sides of the unshaded part are parallel to the three sides of the shaded part respectively. Dissect one shape into the minimum number of pieces and reassemble them to form the other shape.

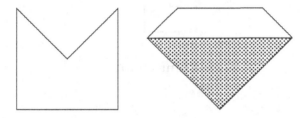

Figure 9.24

Puzzle 24.

Dissect the shape in Figure 9.25 into two pieces and reassemble them to form an equilateral triangle.

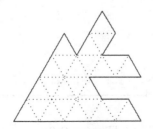

Figure 9.25

Puzzle 25.

Cut a 4 × 9 rectangle into two pieces and reassemble them to form a square.

Puzzle 26.

Dissect the shape in Figure 9.26 into two pieces and reassemble them to form a 4 × 4 square. The shaded square is a hole.

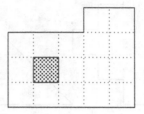

Figure 9.26

Puzzle 27.

Dissect the shape in Figure 9.27 into three pieces and reassemble them to form a square.

Figure 9.27

Puzzle 28.

Dissect the shape in Figure 9.28 into two pieces and reassemble them to form a square.

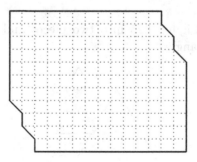

Figure 9.28

Puzzle 29.

Dissect the shape in Figure 9.29 into four congruent pieces and reassemble them to form a square.

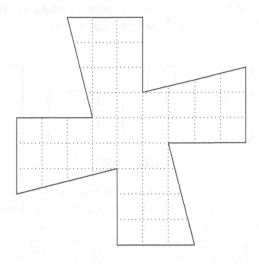

Figure 9.29

Puzzle 30.

Dissect the shape in Figure 30 into three pieces and reassemble them to form a square.

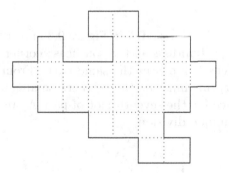

Figure 9.30

Puzzle 1.

Each part consists of four squares joined edge to edge, forming a tetromino. There are five tetrominoes, as shown in Figure 9.31.

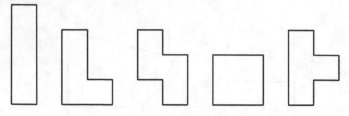

Figure 9.31

Each part may be in the shape of the second or the third tetromino, as shown in Figure 9.32. However, the one on the left does not satisfy the condition that there is a garden in each part.

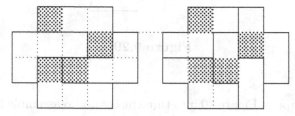

Figure 9.32

Puzzle 2.

The garden and buildings labeled A in Figure 9.33 must belong to the same part. Label the large building at the southeast corner B. Exactly one of the two small gardens belongs in the same part. From the shape of part A determined so far, this must be the northern small garden which belongs to part B. The choice for the seventh plot of part A, and of part B, is now clear, leading to a unique division.

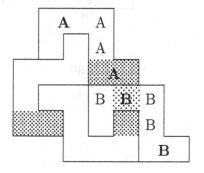

Figure 9.33

Puzzle 3.

Each piece consists of three squares, and is of one of only two possible types, a 1×3 rectangle or a 2×2 square with a corner square missing. The dissections of each of five of the shapes into three pieces of the second type are shown in Figure 9.34. A dissection of the second shape must involve pieces of both types.

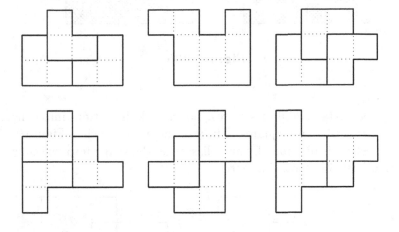

Figure 9.34

Puzzle 4.

The shape consists of 28 unit squares. So we seek dissections into congruent parts consisting of 7 unit squares. Two of them are shown in Figure 9.35.

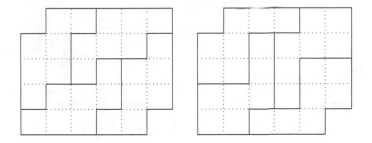

Figure 9.35

Two more dissections are obtained by first dissecting the shape into two congruent pieces that have bilateral symmetry, as shown in Figure 9.36.

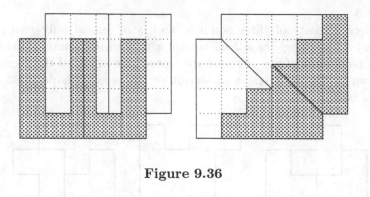

Figure 9.36

Puzzle 5.

The shape consists of 20 unit squares. So we seek dissections into congruent
parts consisting of 5 unit squares. Three of them are shown in Figure 9.37. A
fourth dissection is obtained by first dissecting the shape into two congruent
pieces that have bilateral symmetry.

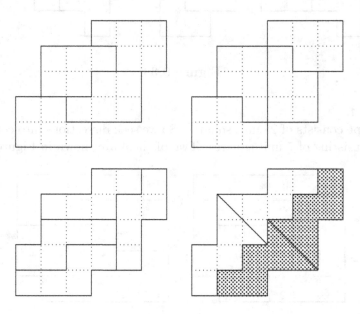

Figure 9.37

Puzzle 6.

The shape consists of 12 unit equilateral triangles. So we seek dissections
into congruent parts consisting of 3 unit equilateral triangles. One of them
is shown in Figure 9.38 on the left. A different kind of solution is shown in
Figure 9.38 on the right.

132

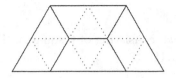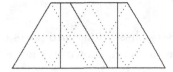

Figure 9.38

Puzzle 7.
Two copies of the given shape form a 6×6 square. Rotating the square 90° yields the desired dissection.

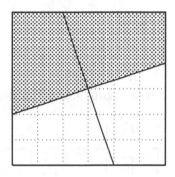

Figure 9.39

Puzzle 8.
Two copies of the given shape form a 6×6 square. Rotating the square 90° yields the desired dissection.

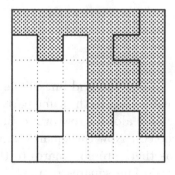

Figure 9.40

Puzzle 9.

(a) We seek dissections into parts each consisting of 6 squares. Two such dissections are shown in Figure 9.41.

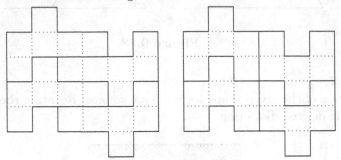

Figure 9.41

(b) Although the dissection in Figure 9.42 is simple, it is not easy to think of it.

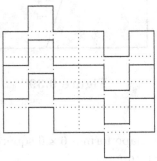

Figure 9.42

Puzzle 10.

(a) We seek a dissection along the grid lines. The seven squares marked A_1 in Figure 9.43 must belong to the same part. Hence there are seven squares in the other part along a row or a column, and those marked B_1 are reasonable choices. Now the squares marked A_2 cannot belong to the second part as there are no squares to the right of the column of squares marked A_1. The squares marked B_2 correspond to those marked A_2. The squares marked A_3 correspond to the squares marked B_3. Now the square marked by a white circle cannot belong to the first part. The square in the first part corresponding to this square is marked with a black circle. The dissection is now clear, and is shown in Figure 9.43.

134

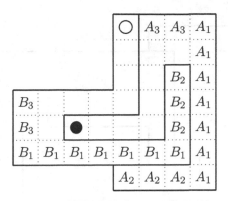

Figure 9.43

(b) First note that this shape can be dissected into two copies of the shape in (a). The dissection in (a) will yield a dissection into four parts here. Second, this shape can be dissected into two congruent pieces in another way, as shown in Figure 9.44. The desired dissection is obtained by cutting along the dashed lines.

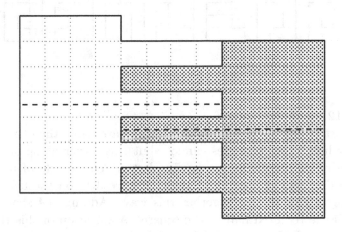

Figure 9.44

Finally, note that the total number of unit squares is 80. Thus we seek dissections into parts each consisting of 20 squares. One such dissection is shown in Figure 9.45.

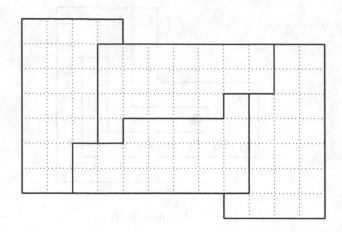

Figure 9.45

Puzzle 11.

There are seven possible dissections.

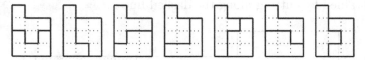

Figure 9.46

Puzzle 12.

The largest triangle will form half of the square while the others will form the other half. Two triangles cannot abut each other along an entire side or an entire diagonal. So we try to abut an entire side of a triangle to an entire diagonal of another triangle. Figure 9.47 on the left shows how five triangles can be put together this way. Adding the shaded triangle completes the required half of the square. A variation of this theme leads to the solution in Figure 9.47 on the right.

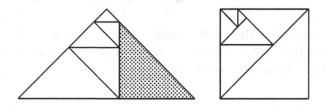

Figure 9.47

Puzzle 13.

We first solve a related problem. Figure 9.48 shows ten circles arranged in two groups of five in two rows. Put one of the numbers 1, 2, 3, 4, 5, 6, 7, 8, 9 and 10 in each circle so that each number in the second row is the sum of the nearest two numbers in the first row.

Figure 9.48

If the number in a doubled circle is odd, the two numbers on each side have opposite parities. If it is even, they have the same parity. Apart from the two numbers in the doubled circles, we have an even number of each parity among the other eight numbers. It follows that the two numbers in the doubled circles have opposite parities. Now the numbers 1 and 2 must appear on the first row while the number 10 must appear in the second row. If the number 9 appears on the first row, it must be next to the number 1, with the number 10 below them, as shown in Figure 9.49.

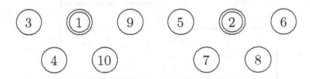

Figure 9.49

Using the first group of numbers as lengths and the second group as widths, we obtain two 13 × 13 squares shown in Figure 9.50.

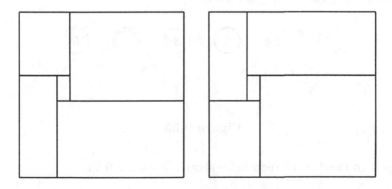

Figure 9.50

137

There is another possibility with 9 in the first row, as shown in Figure 9.51.

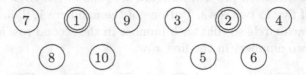

Figure 9.51

This leads to two 9 × 17 rectangles shown in Figure 9.52.

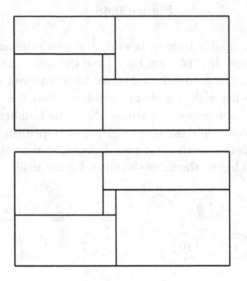

Figure 9.52

Henceforth, both 9 and 10 are on the second row. Suppose they are in the same group, as shown in Figure 9.53.

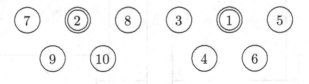

Figure 9.53

This leads to two 9 × 17 rectangles shown in Figure 9.54.

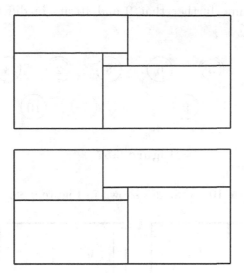

Figure 9.54

There is another possibility with 9 and 10 in the same group, as shown in Figure 9.55.

Figure 9.55

This leads to two 15 × 10 rectangles shown in Figure 9.56.

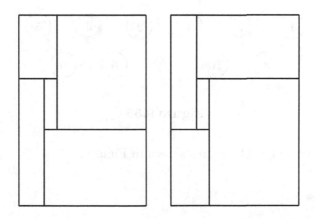

Figure 9.56

Henceforth, we assume further that 9 and 10 are in different groups, as shown in Figure 9.57.

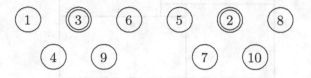

Figure 9.57

This leads to two 15 × 10 rectangles shown in Figure 9.58.

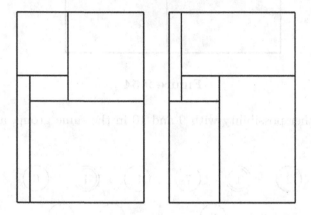

Figure 9.58

There is another possibility with 9 and 10 in different groups, as shown in Figure 9.59.

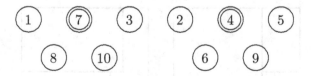

Figure 9.59

This leads to two 11 × 11 squares shown in Figure 9.60.

140

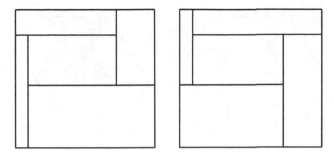

Figure 9.60

Puzzle 14.
If we use the piece in the center which is symmetric, it has essentially only one possible placement. Then the choice and placement of the remaining pieces are forced, leading to the construction in Figure 9.61 on the left. Suppose we do not use this piece. The other symmetric piece is at the northeast corner, and it has essentially two possible placements, as shown in Figure 9.61 in the middle and on the right. In the former case, it is impossible to form the northwest corner of the shape. In the latter case, it is impossible to form the southern edge of the shape. It follows that there is a unique solution up to symmetry.

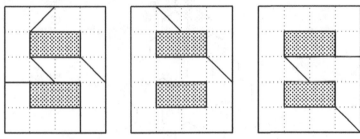

Figure 9.61

Puzzle 15.
If we use the piece at the northwest corner which is symmetric, it has essentially only one possible placement. Then the choice and placement of the remaining pieces are forced, leading to the construction in Figure 9.62 on the left. Suppose we do not use this piece. Another symmetric piece is in the center, and it has essentially two possible placements, as shown in Figure 9.62 on the right. In the position at the bottom, marked by solid lines, two pockets are formed above it. The one on the left is filled as shown, but now the one on the right cannot be filled. In the other position above, which is shaded, it is impossible to form the northeast edge of the hexagon. It follows that there is a unique solution up to symmetry.

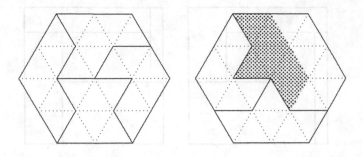

Figure 9.62

Puzzle 16.
When two pieces of opposite orientation come together, a good way is shown in Figure 9.63 at the bottom left corner of the shape. They form a platform on which another piece can rest. It is then easy to fill in the space around them with two more pieces. The bottom right corner can be filled by two pieces of the same orientation forming a sort of parallelogram. The last two pieces, which are of opposite orientations, can fill in the top part of the shape.

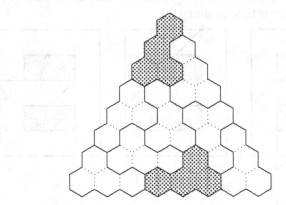

Figure 9.63

Puzzle 17.
Let the side length of $ABCD$ be 1. Then the side length of the target square is $\sqrt{3}$. We can put two of the large pieces together as shown in Figure 9.64, shaded darkly. Their collinear edges, with a total length of 2, will reach from the top of the target square to the bottom at a 60° angle. The third large piece, shaded lightly, can fit in the target square in two ways. It is then easy to place the remaining six pieces.

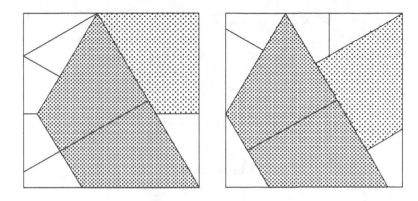

Figure 9.64

Puzzle 18.

When reconstructing the broken scoreboard, there are two positions where the U-shaped piece can be placed, so as to leave room for the 3×2 rectangle. Once it is in place, the positions for the remaining pieces are determined. They are placed so that there are two black circles and two white circles in each column. There are two possibilities, as shown in Figure 9.65. In the former, A and B play C and D on Monday and Saturday, A and D play B and C on Wednesday and Friday, while A and C on Tuesday and Thursday. In the latter, A and B play C and D on Monday and Saturday, A and D play B and C on Thursday and Friday, while A and C on Tuesday and Wednesday. In either case, the winner of the tournament is Team C. Note that the two L-shaped pieces cannot trade places since we must have two black circles and two white circles each day.

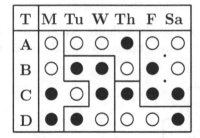 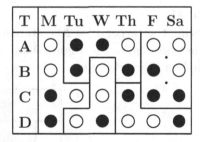

Figure 9.65

Puzzle 19.

Extend the short edges until they meet at the center line from the corner at the bottom, as shown in Figure 9.66 on the left. They cut the figure into three pieces, which can be reassembled into a rectangle, as shown in Figure 9.66 on the right.

143

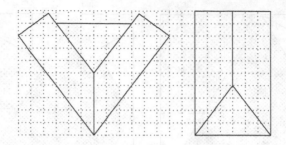

Figure 9.66

Puzzle 20.

Overlaying the target 3×5 rectangle on the given shape, we find that the best position is shown in Figure 9.67. The shaded piece is cut off and moved to complete the rectangle.

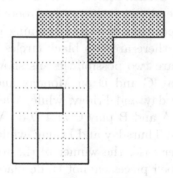

Figure 9.67

Puzzle 21.

Overlaying the target 16×20 rectangle on the given figure, we find that the best position is shown in Figure 9.68. The shaded piece is cut off and moved to complete the rectangle.

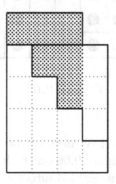

Figure 9.68

Puzzle 22.

We superimpose the two shapes so that the non-overlapping regions are as well connected as possible. One way of doing so is shown in Figure 9.69. A diagonal cut must be made in the non-overlapping part of the first shape. The one shown accomplishes the task.

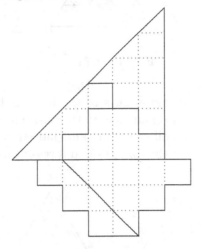

Figure 9.69

Puzzle 23.

Rotate the diamond 135° clockwise, as shown in Figure 9.70 on the left, and superimpose it on the square with the missing quadrant. It is easy to see that we can cut out the darkly shaded part of the diamond and move the piece to the region which is lightly shaded in Figure 9.70 on the right, to complete the square with the missing quadrant. Obviously, the task cannot be accomplished with less than two pieces.

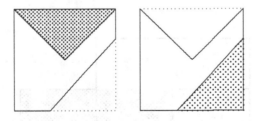

Figure 9.70

Puzzle 24.

The number of unit equilateral triangles is 25, so that the target equilateral triangle has side length 5. In Figure 9.71, we superimpose a copy of it on the original shape. The darkly shaded regions are to be moved to the lightly shaded regions, but the two regions of one kind are not in the same relative positions as that of the two regions of the other kind.

145

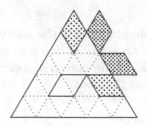

Figure 9.71

If the upper darkly shaded region goes to the lower lightly shaded region, then the lower darkly shaded region goes to the position shown by the blank. This region must then be hollowed out, and moved to the upper lightly shaded region. This leads to the dissection and reassembly shown in Figure 9.72.

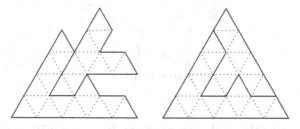

Figure 9.72

Puzzle 25.
Overlaying the target 6 × 6 square on the 4 × 9 rectangle, we find that the best position is shown in Figure 9.73. The shaded piece is cut off and moved to complete the square.

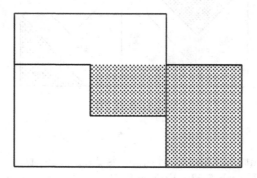

Figure 9.73

Puzzle 26.

Overlaying the target 4×4 square on the given shape, we find that the best position is shown in Figure 9.74. The lightly shaded piece is cut off and moved to complete the square.

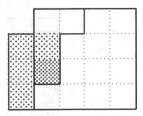

Figure 9.74

A different solution is shown in Figure 9.75.

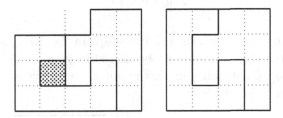

Figure 9.75

Puzzle 27.

The area of the shape is 5. Hence the side length of the target square is $\sqrt{5}$. Now this length occurs as the diagonal of a 1×2 rectangle. Overlaying the target square on the given shape, we find that the best position is shown in Figure 9.76. The two shaded pieces are cut off and moved to complete the square.

Figure 9.76

Puzzle 28.

The area of the shape is 144, so that the square is 12×12. We superimpose it on the original shape, as shown in Figure 9.77 on the left. The part of the shape outside the square is shaded darkly, and this is moved inside the square to become the lightly shaded region.

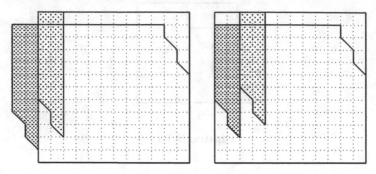

Figure 9.77

Now this region overlaps with the original shape in the darkly shaded part in Figure 9.77 on the right. This must be moved in turn, to the region which is lightly shaded. Continuing in this manner, we arrive at the two-piece dissection shown in Figure 9.78.

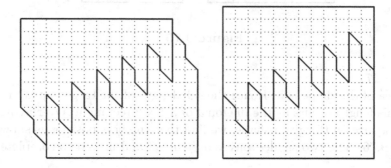

Figure 9.78

Puzzle 29.

As shown in Figure 9.79, copies of the given shape can cover up the entire plane without overlap and without gap. The square has the same property, and a tiling of the plane with copies of a square, equal in area to the given shape, is superimposed on the other tiling. It is shown in dotted lines which pass through the centers of copies of the given shape. The slope of the sides of the squares is either $\frac{4}{5}$ or $-\frac{5}{4}$. This is because the area of the given shape is 41, so that the length of the sides of the squares is $\sqrt{41} = \sqrt{4^2 + 5^2}$.

Figure 9.79

A dissection based on the tilings is shown in Figure 9.80.

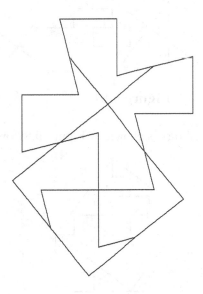

Figure 9.80

Puzzle 30.

Although the given shape consists of 25 unit squares, it is not possible to dissect it along grid lines into three pieces and reassemble them to form a 5×5 square. As shown in Figure 9.81, copies of the given figure can cover up the entire plane without overlap and without gap. The square has the same property, and a tiling of the plane with copies of a square, equal in area to the given shape, is superimposed on the other tiling. It is shown in dotted lines whose slopes are either $\frac{3}{4}$ or $-\frac{4}{3}$. This is because the length of the sides of the squares is $5 = \sqrt{3^2 + 4^2}$.

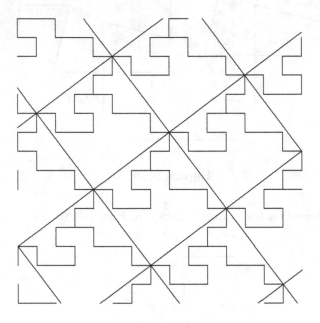

Figure 9.81

A dissection based on the tilings is shown in Figure 9.82 on the right.

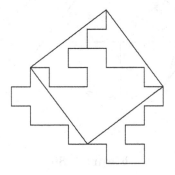

Figure 9.82

150

Chapter Ten
Other Puzzles

Puzzle 1.
The dots in Figure 10.1 appear on a glass door. What do they signify?

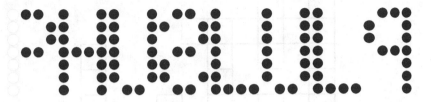

Figure 10.1

Puzzle 2.
Arrange the nine cards in Figure 10.2 to form a four-letter word. The cards may be rotated but not flipped over.

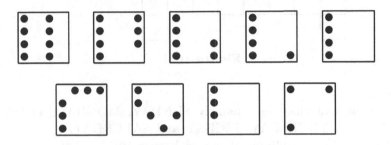

Figure 10.2

Puzzle 3.
Eva, the sponsor of the Young Boy's Association, has composed a word puzzle in which YBA members are asked to fit nine of the words BYE, EVA, FAR, FED, FOB, MID, OAF, OHM, RUE and YBA into the frame shown in Figure 10.3. Each word may be written in either direction. Which word may be omitted?

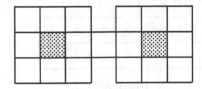

Figure 10.3

151

A. Liu et al., *The Puzzles of Nobuyuki Yoshigahara*, Problem Books in Mathematics, https://doi.org/10.1007/978-3-030-62896-3_10

Puzzle 4.

There are eleven American states with names having at most six letters. Use them to fill in Figure 10.4, with a letter in each box and one name in each row from left to right and in each column from top to bottom.

Figure 10.4

Puzzle 5.

Write the days of the week, namely SUNDAY, MONDAY, TUESDAY, WEDNESDAY, THURSDAY, FRIDAY and SATURDAY, in seven rows. Shift the rows horizontally as shown in Figure 10.5, and the seven-letter name of a Canadian city appears in the shaded column.

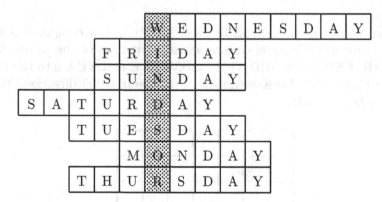

Figure 10.5

Find a major American city with a seven-letter name which can be spelt in this manner.

Puzzle 6.
Freda, Monica, Sarah, Susan, Theresa, Tulia and Wendy are the seven waitresses of the Alphabet Soup Restaurant, each on duty for one day of the week. Business has not been robust of late. To save expenses, the owner decides to close down on Sundays and lay off one waitress. Which one is being laid off?

Puzzle 7.
A woman prefers laughing to crying, watching movies about hijack to those about terrorism, finishing first to finishing second, welcoming refugees from Afghanistan to those from Tajikistan, calmness to noise and defining to deciding. Is she more likely to name her newborn son Arthur or Stuart?

Puzzle 8.
A man prefers visiting New York to Texas, feeding an elk to an elephant, drinking wine to liquor, planting a spruce to an elm, pitching tent in a plain to a valley, collecting artifacts to icons, using pepper to salt, studying electronics to informatics, listening to a robin to an owl, and enjoying sunset to noon. Which is this man more likely to prefer, reading newspapers or watching television?

Puzzle 9.
A boy prefers buccaneers to pirates, chiefs to kings, cowboys to Indians and saints to devils. This boy's favorite color is one of blue, brown and red. Which is it?

Puzzle 10.

(a) Where can one see the string of letters Q W E R T Y U I O P?

(b) Find a ten-letter word which can be spelt using only letters in this string.

Puzzle 11.
James Jason works as a disk jockey for a station with both FM and AM radio. What is remarkable about his name card which reads "J. Jason DJ FM-AM"?

Puzzle 12.
What is shown in Figure 10.6?

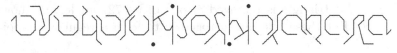

Figure 10.6

Puzzle 13.
A terrorist group announces that a bomb has been placed in the zoo, and will be detonated sometime during July 4. A secret agent is sent to discover the exact time of the attack. He sends back an index card with the cryptic message "ZOOZ". Since it is already known that the zoo is the target, this message seems pointless unless some meaning is conveyed by the final "Z". How should it be interpreted?

Puzzle 14.
A fast driver planning a holiday in Hawaii wants to know the maximum velocity on the autoway. He signs on to his travel agent's website. During login, he comes across the instruction shown in Figure 10.7 on the left. Later, he gets the information shown in Figure 10.7 on the right. What is remarkable about these two displays?

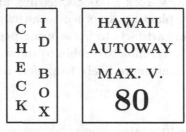

Figure 10.7

Puzzle 15.
In a political convention, the delegates from Ohio are holding placards prepared by their sign printer, who is a poor speller. The state name appears as "OIHO". Fortunately for him, the sign printer is a quick thinker. What action does he recommend which saves his job?

Puzzle 16.
A contractor promises to build a square swimming pool measuring 100 meters from east to west and 100 meters from north to south. When it is completed, its actual area is much less than 10,000 square meters. How can the contractor explain that?

Puzzle 17.
Each of five identical pieces is a rhombus with a 72° angle, and an identical 36° − 72° − 72° triangle is removed at one of its 72° angles, as shown in Figure 10.8. Put these five pieces together to form a regular five-pointed star.

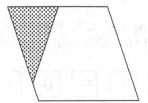

Figure 10.8

Puzzle 18.
Figure 10.9 shows nine matchsticks forming three equilateral triangles. Move
two of the matchsticks to reduce the number of triangles to 0.

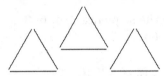

Figure 10.9

Puzzle 19.
Figure 10.10 shows 9 matchsticks forming an expression equal to $2\frac{1}{2}$. Move
only 1 matchstick so that the new expression is equal to 100.

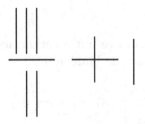

Figure 10.10

Puzzle 20.
Figure 10.11 shows sixteen matchsticks forming the number -101010. Move
only 1 matchstick so that the new expression is equal to nine fifty.

Figure 10.11

155

Puzzle 21.

Seven cards are arranged as shown in Figure 10.12 to form an incorrect equation. Rearrange two of the cards to make a correct equation.

Figure 10.12

Puzzle 22.

What is the value of the product $(x - a)(x - b)(x - c) \cdots (x - y)(x - z)$?

Puzzle 23.

What is the next term of the sequence 2, 4, 6, 30, 32, 34, 36, 40, 42, 44, 46, 50, 52, 54, 56, 60, 62, 64, 66, ... ?

Puzzle 24.

The numbers in Figure 10.13 are put together according to a certain rule. What number should go into the blank circle?

Figure 10.13

Puzzle 25.

The numbers in Figure 10.14 are put together according to a certain rule. What number should go into the blank circle?

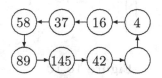

Figure 10.14

Puzzle 26.

(a) The numbers in Figure 10.15 are put together according to a certain rule. What number should go into the blank circle?

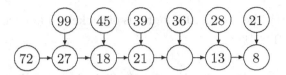

Figure 10.15

(b) The numbers in Figure 10.16 are put together according to a certain rule. What number should go into the blank circle?

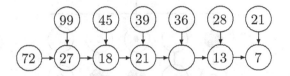

Figure 10.16

Puzzle 27.

Each circle Figure 10.17 represents either a smiling face or a laughing face. Those in the top row are arbitrarily chosen, but all others are determined according to a certain rule. Discover what this rule may be, and determine the kind of faces each of the two blank circles represents.

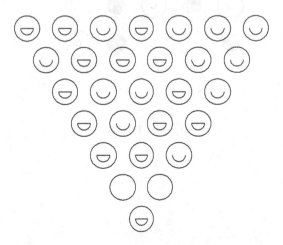

Figure 10.17

Puzzle 28.

An odd number times an odd number is odd, an odd number times an even number is even, but an even number times an even number is odd! Will an odd number plus an even number be odd or even?

Puzzle 29.

Which is worth the most, 10 kilograms of nickels, 5 kilograms of dimes or 2 kilograms of quarters?

Puzzle 30.

A decoration consists of a bar from which five equally spaced strings of five kinds of ornaments are suspended. The bar itself is suspended from its midpoint, and the composition of the strings, as shown in Figure 10.18, keeps the bar horizontal. Ornaments of the same kind have the same weight, while the weight of the connecting pieces of string is negligible. What is the ratio of the weight of a black circle to the weight of a white circle?

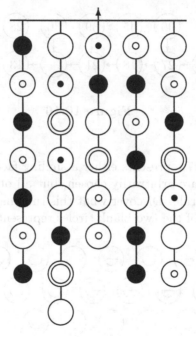

Figure 10.18

Solutions

Puzzle 1.
The dots were printed on both sides of the see-through glass door. Those on the near side are shown in Figure 10.19.

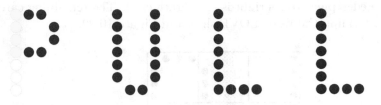

Figure 10.19

Those printed on the far side are shown in Figure 10.20. It is the word "PUSH" reflected.

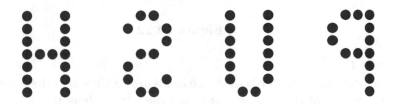

Figure 10.20

Puzzle 2.
Each card is 4×4 so that the overall configuration is 12×12. Most likely, the four letters occupy 5×5 squares at the corners separated by a central cross of width 2. We shade in the potential parts of the cross in the cards as shown in Figure 10.21.

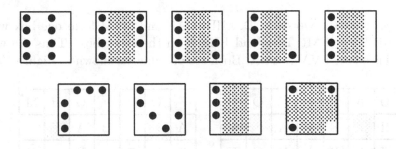

Figure 10.21

It follows that the last card is the central piece while four of the other five cards with shading are the edge pieces. The seventh card is a corner piece, and it can only be part of the letter V. So it is at the southwest corner. The edge piece above it is the fourth card and the edge piece to its right is the third card. Neither the first card nor the sixth card can be at the northwest corner. Hence this corner piece must be the fifth card, forming part of the letter L. The edge piece to its right is the eighth card. The remaining cards fall in place, forming the word LOVE shown in Figure 10.22.

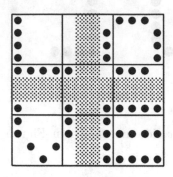

Figure 10.22

Puzzle 3.

We first organize the words as shown in Figure 10.23, where the letters inside the circles are at the ends of words and the letters along the lines are those in between.

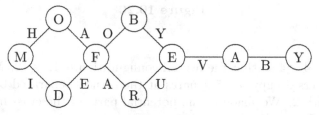

Figure 10.23

It is clear that the words FED, MID, OAF and OHM form one box while the words BYE, FAR, FOB and RUE form the other box. Thus the word omitted is either EVA or YBA. Both are possible, as shown in Figure 10.24.

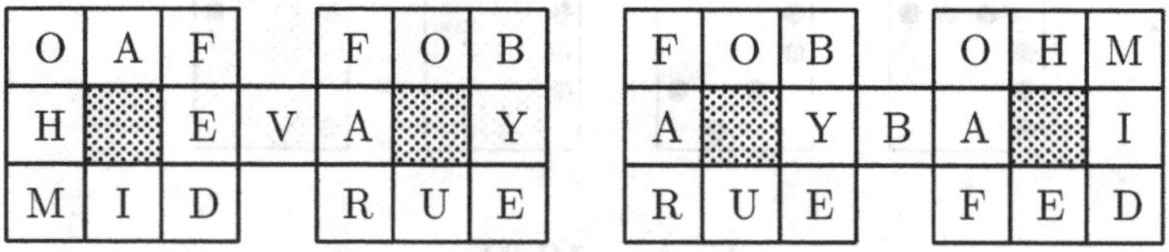

O	A	F		F	O	B			F	O	B		O	H	M
H		E	V	A		Y			A		Y	B	A		I
M	I	D		R	U	E			R	U	E		F	E	D

Figure 10.24

Puzzle 4.

The relevant state names are IOWA, OHIO and UTAH with four letters, IDAHO, MAINE and TEXAS, with five letters, and ALASKA, HAWAII, KANSAS, NEVADA and OREGON with six letters.

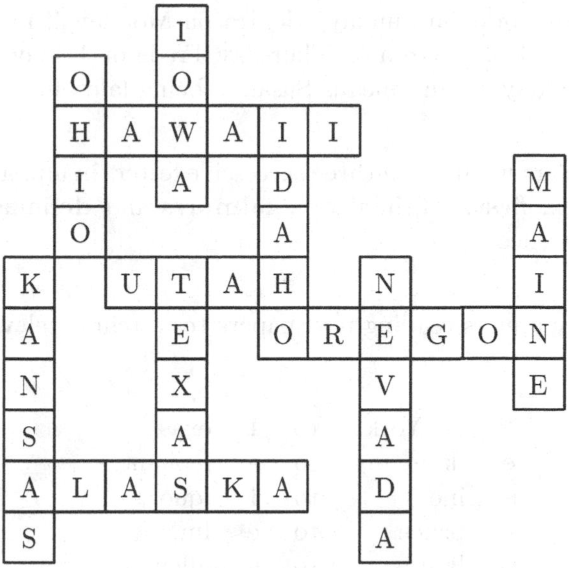

			I							
	O		O							
	H	A	W	A	I	I				
	I		A		D					M
	O				A					A
K		U	T	A	H		N			I
A			E		O	R	E	G	O	N
N			X				V			E
S			A				A			
A	L	A	S	K	A		D			
S							A			

Figure 10.25

Let us first place the state names with four letters. Their second letters are O, H and T respectively. Hence UTAH is the horizontal one, with T being the first letter of TEXAS. The vertical one to the left could be IOWA, with O being the first letter of OREGON. However, the fifth letter of the latter, namely O, cannot be the first letter of another six-letter name. Hence the vertical one to the left must be OHIO, with H being the first letter of HAWAII. The vertical one to the right is IOWA. The rest is easy, as shown in Figure 10.25.

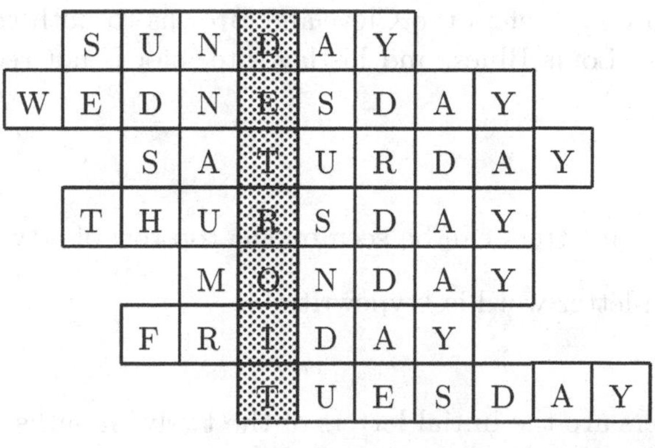

	S	U	N	D	A	Y				
W	E	D	N	E	S	D	A	Y		
		S	A	T	U	R	D	A	Y	
	T	H	U	R	S	D	A	Y		
			M	O	N	D	A	Y		
		F	R	I	D	A	Y			
				T	U	E	S	D	A	Y

Figure 10.26

161

Puzzle 5.

This American city is just across the border, as shown in Figure 10.26.

Puzzle 6.

Apparently, the owner assigns duty to the waitresses according to their names. So Susan works on Sunday, Monica on Monday, Tulia on Tuesday, Wendy on Wednesday, Theresa on Thursday, Freda on Friday and Sarah on Saturday. So Sunday closure means Susan is being laid off.

Puzzle 7.

This woman prefers words with three successive letters in alphabetical order: lau**ghi**ng, **hij**ack, **fir**st, **Afgh**anistan, ca**lmn**ess and de**fin**ing. **Stu**art is clearly her preference.

Puzzle 8.

This man clearly prefers reading newspapers to watching television because he prefers

N	ew York	to	**T**	exas,
e	lk	to	**e**	lephant,
w	ine	to	**l**	iquor,
s	pruce	to	**e**	lm,
p	lain	to	**v**	alley,
a	rtifacts	to	**i**	cons,
p	epper	to	**s**	alt,
e	lectronics	to	**i**	nformatics,
r	obin	to	**o**	wl,
s	unset	to	**n**	oon.

Puzzle 9.

This boy prefers the Tampa Bay Buccaneers to the Pittsburgh Pirates and the Dallas Cowboys to the Cleveland Indians. So he prefers football to baseball. He prefers the Kansas City Chiefs to the Los Angeles Kings and the New Orleans Saints to the New Jersey Devils. So he prefers football to hockey. Therefore he prefers the Cleveland Browns to both the Cincinnati Reds and the St. Louis Blues, and his favorite color is not red or blue but brown.

Puzzle 10.

(a) This string of letters can be seen on the top row of a typewriter.

(b) Such a ten-letter word is "typewriter".

Puzzle 11.

The twelve letters are the initial letters of the twelve months of the year in cyclic order, starting from June.

162

Puzzle 12.
It is an artistic rendering of the name "Nobuyuki Yoshigahara". Nob's name card, designed by the Korean-American genius Scott Kim, reads the same upside down.

Puzzle 13.
Rotating the index card through a right angle, the exact time of attack appears: "NOON".

Puzzle 14.
All letters in the vertical display have horizontal axes of symmetry. All letters and numerals in the horizontal display have vertical axes of symmetry. Moreover, all letters with these properties are used.

Puzzle 15.
He suggests that the delegates should hold the placards upside down.

Puzzle 16.
The contractor creates the impression that the swimming pool is 100 meters by 100 meters, as shown in dotted lines in Figure 10.27.

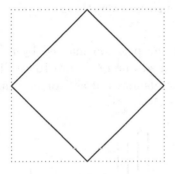

Figure 10.27

The actual swimming pool is obtained by joining the midpoints of adjacent sides of the larger square. Its area is only 5,000 square meters. Yet it measures 100 meters from east to west and 100 meters from north to south.

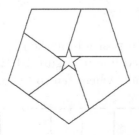

Figure 10.28

163

Puzzle 17.

A regular five-pointed star has 36° angles, but our pieces do not have them. Putting pieces together will only increase the angles. Hence these five pieces cannot form a regular five-pointed star. However, they can form a frame which encloses a regular five-pointed star, as shown in Figure 10.28.

Puzzle 18.

Move the two matchsticks in positions marked by dotted lines to the new positions marked by dashed lines in Figure 10.29. When one triangle is subtracted from one triangle, the difference is equal to zero triangles!

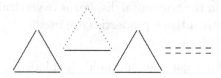

Figure 10.29

Another solution is to strike two matchsticks against each other to set them on fire, and then burn up everything.

Puzzle 19.

Move the matchstick in the position marked by a dotted line to the new position marked by a dashed line in Figure 10.30. The trick is to read the numbers as Arabic numerals and not as Roman numerals. The expression becomes $\frac{1111}{11} - 1 = 101 - 1 = 100$.

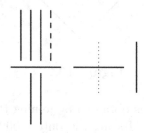

Figure 10.30

Puzzle 20.

Move the matchstick in the position marked by a dotted line to the new position marked by a dashed line in Figure 10.31. The trick is to read the expression as one telling time, where ten to ten is the same as nine fifty.

164

Figure 10.31

Puzzle 21.
We rotate one of the cards with the $+$ sign to make it into a $\times$ sign, and the card with the number 6 to make it into the number 9, as shown in Figure 10.32.

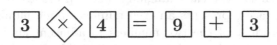

Figure 10.32

Puzzle 22.
Since one of the factors is $x - x = 0$, the product is also equal to 0.

Puzzle 23.
Numerically, one would expect 0, 10, 12, 14, 16, 20, 22, 24, 26, ...to be also included in the sequence. Hence this problem is not numerically based. Let us turn to words. What do "zero", "ten" and "twenty" have in common that is not shared by "thirty", "forty", "fifty" and "sixty"? They all contain the most frequently used letter, namely, "e". If this is indeed the condition for exclusion from the sequence, it agrees with all initial data. Since both "one" and "hundred" contain "e", the next term in the sequence does not appear until we reach "two thousand" or 2000.

Puzzle 24.
The first number is not pointed to by any arrows. Thus it is chosen arbitrarily to get things going. A possible pattern is $49 - 7 \times 7$, $36 = 4 \times 9$ and $18 = 3 \times 6$. So the missing number should be $1 \times 8 = 8$. The pattern cannot continue from here.

Puzzle 25.
A possible pattern is $16 = 4^2$, $37 = 1^2 + 6^2$, $58 = 3^3 + 7^2$, $89 = 5^2 + 8^2$, $145 = 8^2 + 9^2$ and $42 = 1^2 + 4^2 + 5^2$. So the missing number should be $4^2 + 2^2 = 20$. Note that the cyclic pattern does close with $4 = 2^2 + 0^2$.

Puzzle 26.

(a) The numbers in the top row as well as the first number in the bottom row are not pointed to by any arrows. Thus they are chosen arbitrarily to get things going. A possible pattern is $99 - 72 = 27$, $45 - 27 = 18$ and $39 - 18 = 21$. So the missing number should be $36 - 21 = 15$. Note that the pattern does continue with $28 - 15 = 13$ and $21 - 13 = 8$.

(b) Clearly, the pattern here must be different from that in (a). A possible pattern is 27=7+2+9+9, 18=4+5+2+7 and 21=3+9+1+8. So the missing number should be 3+6+2+1=12. Note that the pattern does continue with 12=2+8+1+2 and 7=2+1+1+3.

Puzzle 27.

A possible pattern is that each laughing face represents an odd number and each smiling face represents an even number. Each number below the top row is the product of the two numbers in adjacent circles in the row immediately above. This is consistent with all visible data. Thus the blank circle on the left is a smiling face while the blank circle on the right is a laughing face.

Puzzle 28.

"An odd number times an odd number" has 27 letters, which is an odd number. "An odd number times an even number" has 28 letters, which is an even number. "An even number times an even number" has 29 letters, which is an odd number. "An odd number plus an even number" has 27 letters, which is an odd number. Hence an odd number plus an even number is odd.

Puzzle 29.

If all coins weigh the same, then 10 kilograms of nickels, 5 kilograms of dimes and 2 kilograms of quarters are all worth the same. We need to know that a quarter is heavier than a dime, but a dime is lighter than a nickel. Hence 5 kilograms of dimes is worth the most.

Puzzle 30.

We need to know a little bit of physics. The central string has no effect on the tilting of the bar. Strings to one side tend to pull the bar down to that side. This is called *momentum*, and is clearly directly proportional to the weight on the string. It is also directly proportional to the distance from the string to the center. In fact, the momentum is equal to the product of the weight and the distance. We can verify that the momentum generated by the three kinds of ornaments other than the white circles and the black circles cancel out. All the black circles cancel out except for one on the first string and another on the second string. All the white circles cancel out except for one on the fifth string. Now the black circle on the second string is equivalent to half a black circle on the first string. It follows that one and a half black circles weigh the same as one white circle, so that the ratio of the weight of a black circle to the weight of a white circle is 2:3.

Appendix : Nob and Martin

Nob Yoshigahara and Martin Gardner were and still are central figures in recreational mathematics in their respective cultures. While most of the puzzles in this book come from the creative genius of Nob, there could have been some cross-fertilizations. In this appendix, we list some incidences where identical or similar puzzles have appeared both with Nob and with Martin. Martin has clearly indicated that most of what he writes about comes from other sources. While Nob creates most of his puzzles, he may have been influenced by earlier work, and he has never laid claim to priority.

Nob has written a large number of puzzle books, all in Japanese. We have only scratched at the surface of a mountain trove of treasure. Martin is best known for his monthly *Mathematical Games* column in the magazine *Scientific American*. These columns have been anthologized into fifteen volumes. Currently, there is a joint project between the Mathematical Association of America and the Cambridge University Press to reprint updated editions in uniform format. This is still far from completion. So for our purpose, we use a CD which the MAA has produced, containing the fifteen volumes in their original format. Here is a list of the titles.

1. The Scientific American Book of Mathematical Puzzles and Diversions

2. The Second Scientific American Book of Mathematical Puzzles and Diversions

3. New Mathematical Diversions from Scientific American

4. The Unexpected Hanging and other Mathematical Diversions

5. The Sixth Book of Mathematical Diversions from Scientific American

6. The Mathematical Carnival

7. The Mathematical Magic Show

8. The Mathematical Circus

9. The Magic Numbers of Dr. Matrix

10. Wheels, Life, and other Mathematical Amusements

11. Knotted Doughnuts and other Mathematical Entertainments

12. Time Travel and other Mathematical Bewilderments

13. Penrose Tiles to Trapdoor Ciphers

14. Fractal Music, Hypercards and more Mathematical Recreations

15. The Last Recreations

© The Author(s), under exclusive license to Springer Nature Switzerland AG 2020
A. Liu et al., *The Puzzles of Nobuyuki Yoshigahara*, Problem Books in Mathematics,
https://doi.org/10.1007/978-3-030-62896-3

We now give cross-references between this book of Nob and Martin's columns. We follow the order of the puzzles in Nob's book. In reference to Martin's columns, we use a boldfaced number to indicate the volume (1 to 15 in the above list) followed by page numbers.

Puzzle 1-4

See **4** 239 and 243. Nob's puzzle is based on the same idea as Martin's Question 26.

Puzzle 1-5

See **10** 70 and 79. The two puzzles are essentially the same.

Puzzles 2-1, 2-2, 2-3

See **7** 66–67 and 73, as well as **10** 183–184 and 190, for puzzles with the same theme. Puzzle 2-3 appears in the Postscript of **8** 263–264.

Puzzle 2-5, 2-6

See **6** 108–110 and 118–120 for similar puzzles.

Puzzle 2-7, 2-8

See **1** 112–113 and 120–121 for other examples of matchstick polygons with specific areas.

Puzzle 3-11

See **4** 173–153 and 183 for a similar problem.

Puzzle 4-4

See **3** 135–136 and 142. Nob's puzzle is essentially equivalent to Martin's.

Puzzle 4-7, 4-8

See **2** 35–36 and 39, as well as **6** 13 and 22, for related puzzles.

Puzzle 4-9

See **6** 12–13 and 22. Nob's puzzle is the same as Martin's.

Puzzle 4-11

See **4** 124–125 and 133–135 for related and more complicated puzzles.

Puzzle 4-12

See **2** 214–215. Nob's puzzle is a simpler version of Martin's.

Puzzle 4-14

See **1** 38–39. Nob's game of tic-tac-toe with moving counters is the same as Martin's.

Puzzle 4-15

See **5** 195–196 and 201–203. Nob's game is a simpler version of Martin's.

Puzzle 5-1

See **7** 70 and 77-78. Nob's puzzle and Martin's are identical, special cases of Langford sequences.

Puzzle 5-2

See **4** 76 and 84–85. Nob's puzzle is identical to Martin's.

Puzzle 5-7

See **3** 139 and 147–148 for a similar puzzle.

Puzzle 5-10

See **3** 119–120 and 122. The common theme is coloring maps with three or four colors.

Puzzle 5-11

See **3** 163–164 and 171–172. Nob's puzzle is Latin square whereas Martin's puzzle is a Graeco-Latin square.

Puzzle 5-12

See **10** 189–190 for exactly the same puzzle. It appears again in **13** 268 and 278.

Puzzle 5-15

See **5** 91–93 for other puzzles based on the concept of crossing numbers of graphs. See also **11** 133–144.

Puzzle 5-18

See **3** 156–160. Nob's puzzle appears in Martin's discussion of fault-free rectangles.

Puzzle 5-20

See **4** 201–202. One of Nob's puzzle is the same as Martin's.

Puzzle 7-5

See **2** 138–139. Nob's puzzle is a magic cube, whereas a magic hypercube is featured in Martin's column.

Puzzle 7-7

See **13** 119–120 and 128–129. The theme is triangles of absolute differences or pool-ball triangles.

Puzzle 7-10

See **1** 15–16. Nob's puzzle is a clever variation of Martin's.

Puzzle 7-19

See **10** 14 and 17. Nob's puzzle is exactly the same as Martin's, and Martin attributes it to L. H. Longley-Cook.

Puzzle 7-20

See **10** 12–13 and 17 for a similar puzzle.

Puzzle 8-2

See **9** 65–67 and **10** 153–154. The common theme is the Golomb ruler.

Puzzle 8-3

See **15** 346–347 for exactly the same puzzle.

Puzzle 8-8

See **10** 170–171 and 178. Nob's puzzle is the first of three of Martin's and Martin attributes it to Kobon Fujimura.

Puzzle 8-10

See **14** 155–156 for related puzzles.

Puzzle 8-13

See **13** 269–271 for similar puzzles.

Puzzle 8-14

See **5** 116–118 and 124–126 for similar puzzles.

Puzzle 8-18

See **3** 67–69. The common theme is covering the surface of a cube with straight or bent paper strips.

Puzzle 8-24

See **2** 68 for exactly the same puzzle.

Puzzle 8-25

See **6** 103–103 and 111-113 for related puzzles.

Puzzle 9-6

See **14** 177–179 and 189 for exactly the same puzzle.

Puzzle 9-7, 9-8

See **13** 166–167 and 174–175; in particular, Martin's puzzle 1, 3, 9, 11 and 12.

Puzzle 9-9

See **4** 173 and 182 as well as **10** 21 and 27, for surprise solutions to puzzles similar to Nob's.

Puzzle 9-10

See **13** 179–180. Part (a) of Nob's puzzle appears without solution in the addendum. The puzzle is attributed to Solomon Golomb.

Puzzle 9-13

See **14** 174–175 and 188. Nob's puzzle gives complete solutions.

Puzzle 10-10

See **4** 239 and 243. Nob's puzzle the same as Martin's Question 27.

Puzzle 10-11

See **9** 133. The two puzzles are identical.

Puzzle 10-28

See **4** 240 and 244. Nob's puzzle is based on the same idea as Martin's Question 34.

Printed in the United States
by Baker & Taylor Publisher Services